Acid Mine Drainage: A Threat Posed by Ohio's Historic Coal Mining

Sayeeda

TABLE OF CONTENTS

CHAPTER

1.   INTRODUCTION

1.1 Mining Practices

Acid mine drainage (AMD) is an environmental issue caused by anthropologic activity. Soil and water chemistry is altered due to mining activity, which in turn can change an entire ecosystem. One resource we mine in the United States that leads to this phenomenon is coal. Coal has been mined in Ohio since the early 1800s (Ohio Coal Association, 2015 & ODNR, 2016) and Ohio ranked tenth in the U.S. for coal production in 2010. The first 150 years of coal production were not regulated, and the majority of mining occurring before World War I (1914-1918) was underground mining (ODNR, 2016). These early mines were not typically mapped, and locations were chosen by using vertical, horizontal, and sloping shaft entries to coal seams (ODNR, 2016). The labor was intense, unsafe, and when abandoned they were prone to collapse due to the poor infrastructure during production (ODRN, 2016). As the industry evolved and excavating equipment and explosives became available around the time of WWII (1939-1945), they switched to surface mining and longwall mining (ODNR, 2016).

During surface mining (Figure 1), the soil and rock that is located above the coal seam (known as overburden) is removed and piled away from the current worksite (ODNR, 2016). The exposed coal seam is then mined to remove as much coal-only material possible (ODNR, 2016). Coal that is so intermixed with other materials that it is

deemed unprofitable gets left behind. When mining is completed the work site is

abandoned, leaving behind piles of mine spoil, uneconomical coal, and exposed hillsides

(ODNR, 2016). In the US, AMD and other toxins from abandoned mines have polluted

180,000 acres of reservoirs and lakes and 19,312 km of streams and rivers (Reece, 1995).

Longwall mining is an alteration of underground mining (Figure 1). Instead of

removing the coal by hand, machinery is used to removal the coal in large blocks while

remaining unexposed to the surface (ODNR, 2016). Once the coal is removed there is no

longer support for the overburden, resulting in collapse (ODNR, 2016). The remaining

material, referred to as "gob", is left behind. In the Appalachian region, 72% of mining is

produced this way (EIA, 2016). Our field site located in the Huff Run Watershed does

not have record to determine whether the coal mining in this location was surface mining

or longwall mining activity.

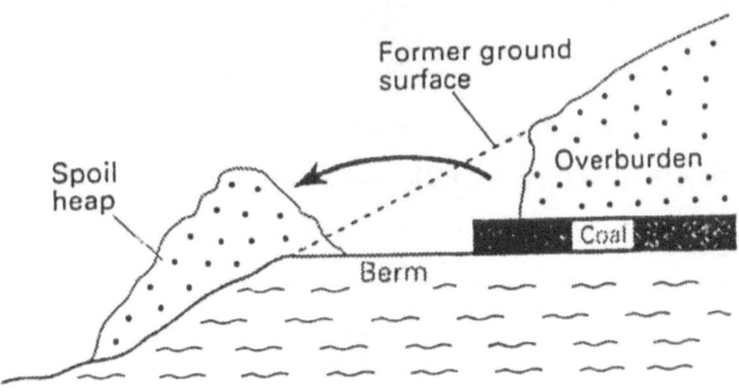

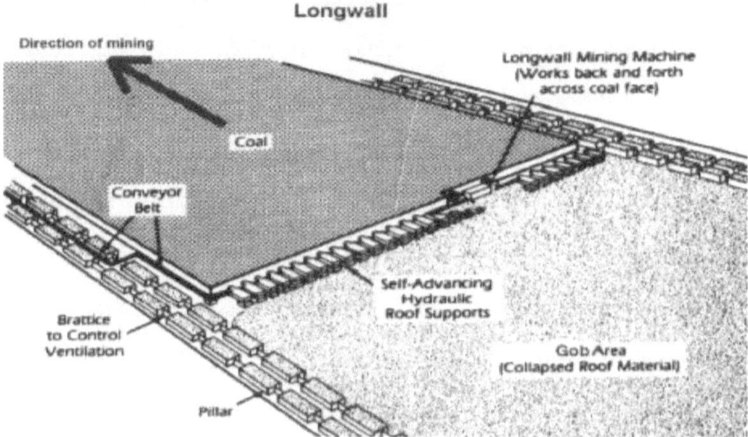

Figure 1: Schematic of surface mining (top) and longwall mining (bottom). Both forms of mining leave behind mine spoil and exposed remnant coal seams. Nalbantov et al. 2010.

1.2 Acid Mine Drainage (AMD)

In the United States, twenty-six of the fifty states are mined for coal, and twenty-four states have United States Geological Survey (USGS) mine drainage projects (USGS, 2006 & Ohio Coal Associate, 2015). These projects study and treat AMD, a consequence stemming from the production of mine waste. Mine waste, or specifically spoil when referring to coal waste, is comprised of topsoil, overburden, surrounding country rock, and uneconomical coal which has been removed and discarded at the surface (Lottermoser, 2010).

The reason that coal poses such issues for waterways is its tendency to contain pyrite (FeS_2). Pyrite, formed under reducing conditions, is stable in the presence of dissolved sulfide and the absence of air (Berner, 1970). When brought to the surface through mining it is no longer stable and undergoes a series of oxidation reactions. The result of these oxidation processes are the release of ferrous iron (Fe^{2+}), sulfate, and hydrogen ions (Reaction 1) (Enviro Sci Inquiry, 2011). The ferrous iron continues to oxidize to form ferric iron and water (Reaction 2). This ferric iron reacts with water to form ferric hydroxide and hydrogen ions (Reaction 3). Pyrite, water, and ferric ions can react to produce ferrous iron, sulfate, and hydrogen ions (Reaction 4).

$$2FeS_2 + 7O_2 + 2H_2O \rightarrow 2Fe^{2+} + 4SO_4^{2-} + 4H^+ \qquad \text{(Reaction 1)}$$

$$4Fe^{2+} + O_2 + 4H^+ \rightarrow 4Fe^{3+} + 2H_2O \qquad \text{(Reaction 2)}$$

$$Fe^{3+} + 3H_2O \;\rightarrow\; Fe(OH)_3 + 3H^+ \qquad\qquad \text{(Reaction 3)}$$

$$FeS_2 + 14Fe^{3+} + 8H_2O \;\rightarrow\; 15Fe^{2+} + 2SO_4^{2-} + 16H^+ \qquad\qquad \text{(Reaction 4)}$$

These chemical reactions can reduce the pH and has been documented in some cases to be less than 2 (Strawn et al., 2015). As water passes through these soils it can flow into water systems such as streams, which in high enough volume changes the pH, and increases its metal concentrations.

This phenomenon is known as acid mine drainage (AMD), and with its negative impact on ecosystems has long lasting impacts on water quality and stream ecology in affected watersheds, and has been the fuel to begin remediation projects all across the United States. The remediation of environmental damage caused by these mines is extremely costly. Between 2005 and 2012, monitoring and reclamation of over 300 km of streams and rivers in Ohio was done at a cost of over $25 million dollars (Bowman and Johnson, 2012).

Current studies on AMD tend to address point sources and their impact on streams. A point source, as defined by the EPA, is a source of pollution that is "discernable, confined, and of discrete conveyance", such as a pipe outlet (EPA, 2011). The EPA describes a nonpoint source as "pollution generally results from land runoff, precipitation, atmospheric deposition, drainage, seepage or hydrologic modification" (EPA, 2011). Precipitation drives nonpoint source pollution by washing over the pollutant and carrying it over the land and into water systems (EPA, 2011).

Although there is extensive literature on reclaimed and remediated coal mine tailings (for review, see (Hossner and Hons, 1992; Johnson and Hallberg, 2005)), few studies have addressed the fate of trace metals in historic mine tailings (stemming from the non-point sources), which were abandoned in the early 20th century and now blend into the surrounding landscape. These topographic highs are known to be long-term sources of slow-leaching AMD adjacent to reclamation projects that can hinder effective remediation (Wise, 2005).

1.3 Project Objective

In this study, two hills of similar size and elevations from the Huff Run Watershed were sampled. One was a native hill, and the other was a hill formed from mine spoil. This location was chosen so that our data can be easily shared with the Huff Run Reclamation Projects. Data collection and remediation projects have taken place over the last 45 years, though several organizations, and they have reached their goal of bringing all 15.9 km of the stream up to a pH of 6.5 (Kinney, 2013). We are hoping that our data will contribute to maintaining the healthy pH of the waterhshed, and offering explanation for areas that may not continue to produce healthy results in the long term as expected.

2.   SITE DESCRIPTION

2.1 Coal Formation

Coal, a sedimentary rock, is formed under reducing conditions where organic matter accumulates faster than it can be fully oxidized or break down (Babcock, 2009). It requires a large carbon source, such as a swamp, which after burial undergoes heat and compaction. A common time in geologic history for coal deposit formation is the Carboniferous Period (UCMP, 2009).

Carroll and Tuscarawas counties are south of the Glacial Margin in Ohio and have bedrock dating back to the Carboniferous Period, specifically the Pennsylvanian. During the Carboniferous Period, sediments were sometimes deposited as cyclothems. These deposits represent a series of beds, commonly including coal beds, that typically represent a transgressive-regressive sequence (Babcock, 2009). The glacial-interglacial cycles of the Carboniferous Period provided the sea level change for these cyclic beds (Babcock, 2009). Miners working in areas of cyclothems find that deposits end up mixing the different depositional beds together, resulting in a loose mix of sandstone, shale, limestone, and coal (Babcock, 2009).

2.2 Sampling Location

The Huff Run watershed (Figure 2), located in both Carroll and Tuscarawas counties, has sediments from the swampy timeframe of the Carboniferous, and is dominated by silt loams per the Web Soil Survey (Appendix 1). This watershed encompasses 36.5 km^2 of land with a 15.9-km stream running from Carroll to Tuscarawas county (Ohio) and empties into Conotton Creek, South of Mineral City (southwesterly flow) (Fleming, 2010). The land is approved for use as a warmwater habitat, primary contact, and public water supply for agricultural and industrial use (Fleming, 2010). This watershed has around 30% of its land affected by mining operations and AMD (EPA, 2005).

Our sampling location is designated as HR-25, a 0.7 km tributary which is located in Reach 2, which ranked as the number 1 problem site for water quality problems in this watershed, with a metal (Fe and Al) loads value of 28.3 kg/d (Fleming, 2010). The source for the AMD in this location is from the Strong Mine which was abandoned in 1922 (Kinney, 2013). Although water in the primary stream channel is currently alkaline, the tributary of study, along with several others are still acidic, and can lower the pH of Huff Run, or overwork remediation efforts (Kinney, 2013). It is estimated that "if HR-33, HR-36, HR-25 and Farr were effectively treated approximately 58% of the acid load to Huff Run would be eliminated under low flow, 61% under medium flow, and 40% under high flow" (Kinney, 2013). Successful remediation of HR-25 would aid in recovering the lower 3.7 river km of the watershed (Kinney, 2013).

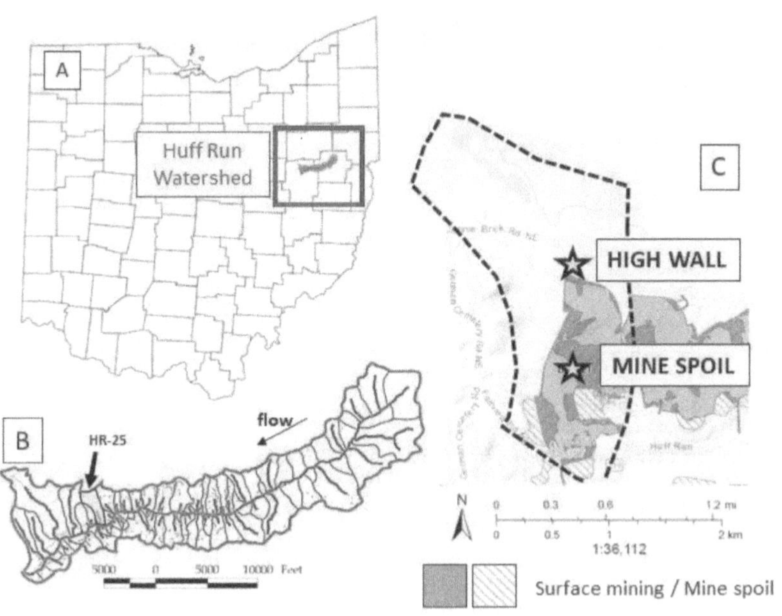

Figure 2: Map of Huff Run Watershed. (A) Location of Huff Run Watershed within the state of Ohio; (B) Huff Run Watershed with sampling site (HR-25 highlighted); (C) spatial extent of known surface area within HR-25 (the dashed outline) affected by coal surface mining (green) and mine spoil emplacement (hatched orange), modified after (Singer et al., 2021b). The stars mark the locations of the high wall and mine spoil sites. Singer et al., 2021

2.3 Huff Run Watershed Restoration Partnership

To combat the effects of AMD The Huff Run Watershed Restoration Partnership
(HRWRP) was formed in 1996 and has completed 20 reclamation projects to date
(Baker, 2015). The total cost of these projects totals around $6 million, primarily
funded using coal excise taxes (Baker, 2015). These reclamation projects have
performed tasks such as installing limestone channels and steel slag beds, re-grading
and capping mine tailings, filling acid pits, and surface reclamation (NPS, 2014). The
HRWRP is still receiving funding, both from coal taxes and the state, and it will
continue to remediate as long as money is provided (Baker, 2015).

A 2014 NPS report shows that the efforts of HRWRP have benefited the area
(Figure 3), but with so many non-point source contamination points available the
longevity of these projects are questionable. There has also been fluxes of metal and
acid loads to the river, such as in 2016, where the river temporarily did not meet its
pH goal (NPS, 2017 – 2018). The Huff Run Watershed follows the Acid Mine
Drainage Abatement and Treatment Plan (AMDAT) that was provided by the Ohio
Department of Natural Resources (ODNR) (Fleming, 2010). In this plan the goals
include, "identification of the hydraulic unit; determination of current conditions and
the restoration potential of the watershed; establishment for goals for restoration;
definition and prioritization of potential projects for funding; and recommendations
for future monitoring programs".

Although the HRWRP has continually referenced these goals, reclamation projects do not have an infinite longevity. They can become exhausted, breakdown, or stop functioning as intended for a variety of reasons. For this reason, the Huff Run Watershed has future monitoring programs in place, so they will be able to ensure the reclamation projects are still sound. If the pH drops back down to prior reclamation levels our research will contribute to formatting a plan on whether further investments should be made in this watershed.

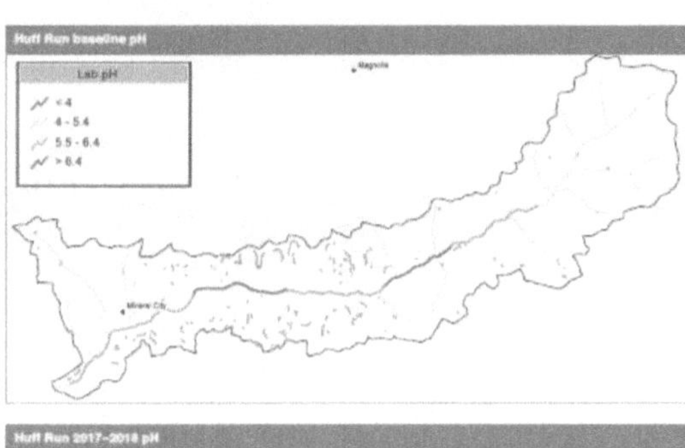

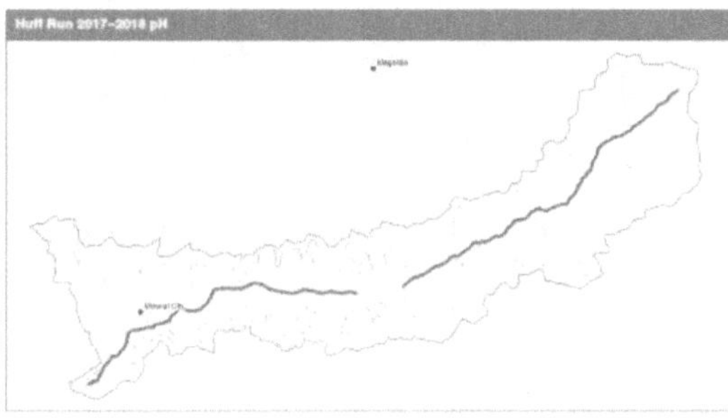

Figure 3: Huff Run pH map. A 2017 – 2018 NPS report shows that the stream pH baseline generated from data gathered from 1985-1998 has improved to the targeted pH of 6.5 since 1999. There is a missing section on the 2017 – 2018 map due to lack of monitoring data. NPS, 2018

Even though the stream has attained the pH goals, there are still many areas of the

watershed that are overloaded with AMD. Walking through various areas of the

watershed in HR-25 you will not be hard pressed to find orange soil banks and water

along ditches and minor creeks (Figure 4). The HRWRP has carefully chosen the

locations of all their remediation projects to meet their pH and metal loading goals for

the primary stream in the watershed. Without the use of substantially more

resources, it is impractical to remediate an entire watershed, so having non-point

source data from a highly contaminated area such as HR-25 can be valuable in

making future remediation decisions.

Figure 4: Pictures taken throughout HR-25 showing orange water and soil banks due to
high metal loading. Photos by Laura Zemanek

3.   METHODS

3.1 SOLID PHASE METHODS

3.1.1 Sampling Location

Two hills were chosen for this study based upon their similarity in elevation, vegetation, and proximity (Appendix 2). A soil core was taken from a vegetated high wall (HW), and a vegetated mine spoil hill (MS) to a depth of 120 centimeters. The HW is located at 40°36'20"N, 81°20"40"W, at an elevation of 326.1 meters. MT is located at 40°36'17"N, 81°20'40"W, at an elevation of 304.8 meters. Soil cores were taken at the peak of each hill to eliminate the mixing of soil from more than one location on the hill. Tree cores were taken from a blue beech tree growing on the hilltop of the mine spoil site. It was chosen based on its larger size in comparison to other trees on the hill. To roughly estimate how long the tree has been growing, its rings were counted giving an age no greater than 50 years. Using this information, it is likely that the mine soil which it has grown upon has been undisturbed for the past 50 years.

3.1.2 Soil Collection and Division

A steel AMS hand auger, with a diameter of 5.715 cm, was used to pull up soil in increments of 10 centimeters at a time, and the samples were bagged in plastic. Color and texture changes in the soil profile were noted for later placement of lysimeters. The soil was then placed into aluminum pans and dried in an oven (Binder Oven and a Fisher Scienfitic Isotemp Oven) at 60°C for 24 hours. After drying, the samples were weighed (Appendix 3) and divided into four equal fractions using a Humboldt riffle type splitter, and labeled as A, B, C and D. Fraction A was milled into a powder using a SPEX SamplePrep 8000 M Mixer/Mill for 6 minutes. This powder was used for loss on ignition and XRD analysis. Fraction B was broken down into finer particles using a ceramic mortar and pestle until the material was able to be sieved using a Retsch® Test Sieve ASTM E11 63 µm. This sieved sample was used for SEM analyses. Fraction C was separated by particles size above and below 2mm using a Retsch® Test Sieve ASTM E11 2 mm (mortar and pestle used to break apart aggregates) and used for particle size analysis (PSA). Fraction D was archived.

3.1.3 Particle Size Analysis (PSA)

The particle size distribution from each soil sample was determined with a Retsch Camsizer Video Particle Size Analyzer (for the 1 mm – 2 mm fraction), and a Malvern Mastersizer 2000 (for the less than 1 mm size fraction). After drying, soil samples were passed through a 2-mm sieve to remove larger particles, and then passed through a 1-mm sieve. The mass of the 1 – 2 mm sample was weighed in order to be summed with the post-1mm sieved sand size fraction results. The sieved powders were suspended in deionized water. The size fraction for sand, silt, and clay were defined as > 63 µm, 2 – 63 µm, and < 2 µm, respectively.

3.1.4 X-ray Powder Diffraction (XRD)

X-ray diffraction patterns were collected on powdered samples from fraction A using a Rigaku Geigerflex x-ray diffractometer (3-70°, 0.02; 2.0° step size; 2/min). Glass slides were washed with acetone between samples. Phase identification was performed using JADE 6.5 (Materials Data Inc., Livermore, CA). Using the initial peak identification, a quantitative phase analysis was performed the Rietveld module included in the X'Pert HighScore Plus software (Perdrial et al., 2011). Diffractograms from XRD were produced and sent to Nicholas Perdrial of the University of Vermont for a quantitative analysis (Appendix 4).

3.1.5 Loss On Ignition (LOI)

One gram of soil was taken from fraction A, placed into ceramic crucibles and heated in a Thermo Scientific Thermodyne muffle furnace at 550 °C for four hours to determine the mass of organic matter contained in the soil, known as loss on ignition procedure. Afterwards, the crucibles were left to cool for 2 hours, then reweighed to determine the amount of organic matter lost. This process was performed in triplicates and graphed as weight percent lost. Three samples were then run at a temperature 950 °C for the depths of 40-50 cm, 90-100 cm, and 110-120 cm (depths chosen for most significant color change in powdered form).

3.2 AQUEOUS PHASE METHODS

3.2.1 Lysimeter Installation

Lysimeters (1900 Soil Water samplers, Soil Moisture Corp.), were installed in the mine spoil and high wall hill on May 13, 2015, following soil collection (Figure 5). Depths were chosen to represent different soil horizons based on color changes seen in the MS; lysimeters were placed at 10 (organic horizon which transitioned to a more orange color from 20-30 cm), 40 (very clayey and difficult to remove from auger), 80 (dark material which was soft and squishy), and 120 cm. The high wall profile was difficult to decipher any horizon changes, and overall was a darker browner color, therefore the lysimeter placement was solely based on the horizons found in the MS hill. The base of the holes were filled with silica flouer, wetted until the consistency was that of cement mortar, to ensure that theceramic cup would have a clean hydraulic contact without bias from the surrounding soil (Soilmoisture, 2007). The lysimeter was then placed into the silica slurry, with more silica flouer added to exceed the height of the ceramic cup, and bentonite clay pellets were added on top of the silicia to seal the ceramic cup from vertical flow paths from the area above it. The hole around the length of the lysimeters was then backfilled with soil that was sieved, with an ASTM E-11 standard ½ inch sieve to remove pebbles and rocks. The soil was packed firmly around the lysimeters using a meter stick, and another bentonite seal was poured around the lysimeters near the surface to avoid rainwater contamination. Soil was used to cover the bentonite, and slightly mounded and tampered down along the top of the lysimeter.

Suction was put on the lysimeter using a 2005G2 Vacuum Hang Pump at a pressure of 60

kPa to create a negative pressure inside the soil water sampler (Figure 5).

Figure 5: Lysimeters after installation, located at the high wall. The
dashed line represents the edge of the hill where the miners stopped
excavating a coal seam.

3.2.2 Lysimeter Water Extraction

A 60 mL syringe connected with a three-way valve to long plastic tubing was inserted into the lysimeter to pull water out from the ceramic cup of the lysimeter. The first 5 mL of sample was used to rinse the syringe and saturate the filter, and then discarded before sampling. During sampling, pH, dissolved oxygen, electrical conductivity, and temperature were measured in the field using a HACH Waterproof Handheld H160, Pro1020 YSI, and HM Digital aquapro water tester, respectively. Field tests were later averaged for each depth, and standard error was calculated. Water was then filtered using a Target 2 nylon 0.45 μm filter into three separate vials for each given sampling depth. The three vials were used for sampling: (A) metals for analysis by inductively coupled plasma atomic emission spectroscopy (ICP-OES) using acid-washed plastic vials and acidified with 1-2 drops of concentrated nitric acid (~70%); (B) anions for analysis by ion chromatography using acid-washed plastic vials and not acidified; and (C) dissolved organic carbon (DOC) using combusted glass vials (caps were rinsed with Milli-Q, Q-POD Millipore 0.22μm water) and 1-2 drops of concentrated hydrochloric acid (~35%). Residual water remaining in the lysimeter was removed after sampling prior to putting the lysimeter under vacuum. The lysimeter installed in the MS between 110 and 120 cm was frequently dry. Dates of collection with field results are found Appendix 5.

3.2.3 Inductively Coupled Plasma Optical Emission Spectrometry (ICP-OES)

Aqueous ion concentrations in soil pore water (vials A) were determined using a
PerkinElmer ICP-OES 8000 for the following elements: Al, As, Cu, Fe, Mn, Ni, Se, and
Zn. A standard of 1 mg/L Mn solution was run before the samples were analyzed and a
2% nitric acid rinse was run between samples and standards. Results were collected in
analytical triplicates using WinLab for ICP software (EPA method 2007).

3.2.4 Ion Chromatography (IC)

Anions in the soil pore water (vials B) were analyzed using a Thermo Scientific
Ion Chromatography (IC) system. Samples were run at full concentration and at a 1:20
dilution so that results were between 0.1-50 ppm, and obtained results for fluoride,
chloride, nitrite, bromide, sulfate, and phosphate. A list of IC values and calibration
curves can be found in Appendix 6.

3.2.5 Dissolved Organic Carbon (DOC)

Dissolved organic carbon, (vials C) was analyzed using a Shimadzu Scientific
Total Organic Carbon (TOC). Samples taken from the high wall at a depth of 10 cm
were (high in dissolved organic content which exceeded the upper limit of 50 mg/L
carbon; water was straw yellow in color, while all other locations were clear in color)
diluted at a 1:10 factor (Appendix 7).

3.2.6 Sequential Extraction

Sequential extraction, modified after Tessier (1979), was done for each sampling depth, with triplicates performed at depths of lysimeter installation. Powdered soil (1 ± 0.052 g) was agitated with four sequential solutions to determine element distribution in operationally defined fractions. After agitation with each solution, the sample was centrifuged at 4,000 rpm for 30 minutes and decanted. 10 mL of Milli-Q water was then added and centrifuged for an additional 30 minutes and decanted to remove and remaining solution and ions before adding the next extraction solution. Solution 1 *Exchangeable*: 8 mL - 2M NaCl solution at room temperature, pH 7, was agitated with the soil for 2 hours; Solution 2 *Bound to Carbonates*: 8 mL - 1 M sodium acetate (buffered to pH 5 by adding acetic acid) agitated in solution for 5 hours at room temperature; Solution 3 *Reducible*: 20 mL - 0.3 M $Na_2S_2O_4$ (Sodium Dithionite) + 0.175 M Na-citrate + 0.025 M H-citrate agitated in solution for 6 hours at room temperature; Solution 4 *Oxidizable*: 3 mL - 0.02 M HNO_3 + 5 mL 30% H_2O_2, adjusted to pH 2 by adding additional H_2O_2, was heated to a temperature of 85°C for 2 hours, with occasional agitation. 3 more mL of 30% H_2O_2 was added, with pH adjusted to pH 2 by adding HNO_3, was then heated for 3 hours with occasional agitation. After cooling, 5 mL of 3.2 M NH_4OAc in 20% (v/v) HOAc was added, with Milli-Q water added to reach 20 mL if needed. The extracted solution along with the rinse solution was analyzed by ICP-OES for Fe, Mn, Al, Cu, Ni, As, and Se.

4.   RESULTS

4.1 SOLID PHASE RESULTS

4.1.1 Particle Size Analysis (PSA):

Particle size distribution ranged from 3.2 - 4.8% clay, 37.6 - 54.1% silt, and 41.2 - 59.0% sand for the HW, and 5.1 - 7.4 % clay, 35.7 – 47.8% silt, and 45.0 – 58.7% sand in the MS (Figure 6). The average compositions were 4.4% clay, 47.7% silt, and 47.9% sand for the HW, and 6.3% clay, 41.8% silt, and 51.9% sand for the MS. The sand slightly increased its proportion with depth, while there was no significant trend of particle size in the HW. The MS had slightly more clay content along its entire profile than the HW (~2%). Values for each depth are in the Appendix 8.

4.1.2 X-ray Powder Diffraction (XRD):

Both soils were dominated by quartz and muscovite, made with similar proportions, and in total comprising 76.2% of the HW profile, and 72.7% of the MS profile. In the HW, there was on average 2.3% more muscovite than quartz. For both profiles, the remaining mineral averages follow in the order of most abundant to least abundant with kaolinite (5.0% more in MS), feldspar (1.6% more in HW), and then chlorite (<0.1% more in MS) (Figure 11).

The total HW mineral composition ranged from 30.6 – 41.0% quartz, 34.9 – 42.6% muscovite, 12.3 – 15.5% kaolinite, 3.9 – 8.7% feldspar (the sum of anorthite, albite, and orthoclase), and 2.1 – 4.0% chlorite. The MS mineral composition had more variance for each mineral and ranged from 27.5 – 49.0% quartz, 29.1 - 45% muscovite, 16.1 – 22.4% kaolinite, 1.1 – 9.4% feldspar, and 0.8 – 7.8% chlorite (Figure 6). The average composition for the HW was $37.0 \pm 3.3\%$ quartz, $39.2 \pm 2.4\%$ muscovite, $14.3 \pm 1.0\%$ kaolinite, $6.4 \pm 1.3\%$ felspars, and $3.2 \pm 0.6\%$ chlorite. The average composition for the MS was $36.6 \pm 6.5\%$ quartz, $36.1 \pm 4.3\%$ muscovite, $19.3 \pm 1.9\%$ kaolinite, $4.8 \pm 2.0\%$ felspars, and $3.3 \pm 1.9\%$ chlorite. There is more kaolinite and less feldspar in the MS, and when summing the average content of chlorite, muscovite, and kaolinite the MS has 2.03% more of these clay size minerals throughout the profile (HW ~58.7%, MS ~56.7%). This 2% matches the 2% increase of clay particles in the MS found during PSA. A table with values for each depth can be found in Appendix 9.

4.1.3 Loss on Ignition (LOI)

The HW LOI values exhibited little variance between depths with readings of 4.27 ± 0.70 to $8.53 \pm 4.55\%$, whereas the MS soil showed high variance with values of 5.35 ± 1.53 to $15.48 \pm 2.67\%$ (Figure 6). The average LOI for the HW was $7.02 \pm 1.95\%$ and the average LOI for the MS was $9.49 \pm 1.28\%$. Both sites had their lowest percent lost at 30-40 cm, with the MS at 5.35% and the HW at 4.27%. They also had their highest percent lost at a depth of 50-60 cm, with the MS at 15.48% and the HW at 8.5%.

There were only two depths where the LOI was higher in the HW. These depths include

20-30 cm (MS = 5.75% vs. HW = 7.15%) and 40-50 cm (MS = 5.93% vs. HW = 6.21%).

Depths of 10-50 cm had low LOI values, with its lowest point at 30-40 cm with an LOI

of 5.35%. Intermediate values were found in the initial 0-10 cm and 70-120 cm, and a

high LOI of 15.48% was found at 50-60 cm of depth (Figure 11).

The three samples for each location (40-50 cm, 90-100 cm, and 110-120 cm) run

at a temperature of 950°C had a percentage lost when comparing the HW to the MS was

less than 0.5%. For the three sample depths run at an increased temperature (950 °C)

resulted in an additional 2.89-3.81% loss, with only a 0.5% difference when comparing

the HW to the MS. A table with values for each depth can be found in Appendix 10.

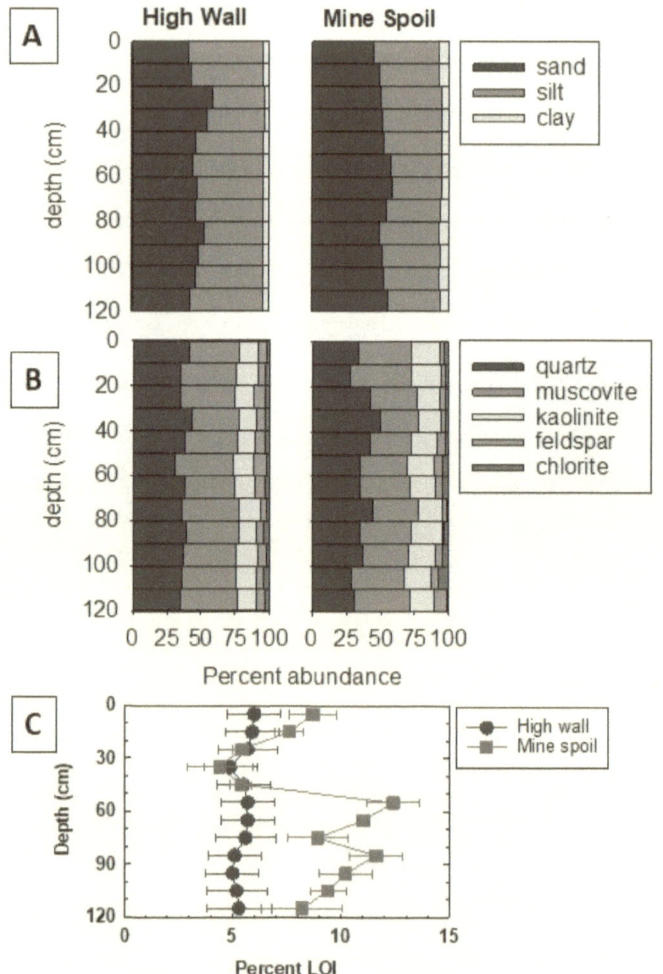

Figure 6: Comparison of High Wall (left) and Mine Spoil (right) particle size analysis (A), mineralogical composition (B); the bottom panel (C) shows loss-on-ignition (LOI). Singer et al., 2021

4.2 AQUEOUS PHASE RESULTS

4.2.1 pH:

The average pH for each depth was calculated to compare the HW to the MS
(Figure 7). The pH of the HW samples ranged from 5.49 – 7.37, with an average value of
6.45 ± 1.61 at 10 cm, 5.93 ± 1.58 at 40 cm, 6.08 ± 1.75 at 80 cm, and 6.24 ± 1.51 at 120
cm. The pH of the MS samples ranged from 4.13 – 7.07, with an average value of 6.29 ±
2.10 at 10 cm, 5.43 ± 1.40 at 40 cm, 5.40 ± 1.31 at 80 cm, and 5.77 ± 1.74 at 120 cm.

4.2.2 Dissolved Oxygen (DO):

The average DO as a function of depth for the HW was as follows: 5.79 ± 1.75 at
10 cm, 4.65 ± 1.29 at 40 cm, 4.50 ± 1.42 at 80 cm, and 4.79 ± 1.24 at 120 cm. The
average DO for the MS samples was: 4.69 ± 2.71 at 10 cm, 4.50 ± 1.30 at 40 cm, 3.46 ±
0.89 at 80 cm, and 4.08 ± 1.82 at 120 cm. On July 28, 2015 both the high wall and mine
spoil had their lowest value of DO, with the high wall at 2.92 mg/L and the mine spoil
with 2.00 mg/L, both of these values were recorded at 80 cm of depth (Appendix 5). The
highest DO value for the HW was 8.70 mg/L at 10 cm, and the highest value for the MS
was 6.67 mg/L at 40 cm (Figure 7)

4.2.3 Electrical Conductivity (EC):

The average EC as a function of depth for the HW are as follows: 743.31 ±
185.83 µS/cm at 10 cm, 764.86 ± 204.42 at 40 cm, 789.83 ± 228.01 at 40 cm, 789.83 ±
228.01 at 80 cm, and 798.24 ± 193.60 at 120. The average EC for the MS samples was
766.67 ± 255.56 µS/cm at 10 cm, 381.60 ± 98.53 at 40 cm, 450.53 ± 109.27 at 80 cm,
and 791.45 ± 238.63 at 120 cm. All average readings for both the HW and MS were
within 55 µS/cm of each other, with exception of MS depths of 40 and 80 cm (Figure 7).

4.2.4 Temperature (°C):

The average temperatures as a function of depth for the HW were as follows: 21.2
± 5.47 °C at 10 cm, 20.2 ± 5.59 at 40 cm, 20.3 ± 6.13 at 80 cm, and 20.3 ± 5.07 at 120
cm. The average temperatures as a function of depth for the MS were as follows: 21.4 ±
7.57 °C at 10 cm, 20.7 ± 5.52 at 40 cm, 20.3 ± 5.07 at 80 cm, and 20.6 ± 6.51 at 120 cm.
The temperature for all the sampling locations was 15.1°C or less on May 20th, 2015,
with early June through early August reflective of the average values (Figure 7).

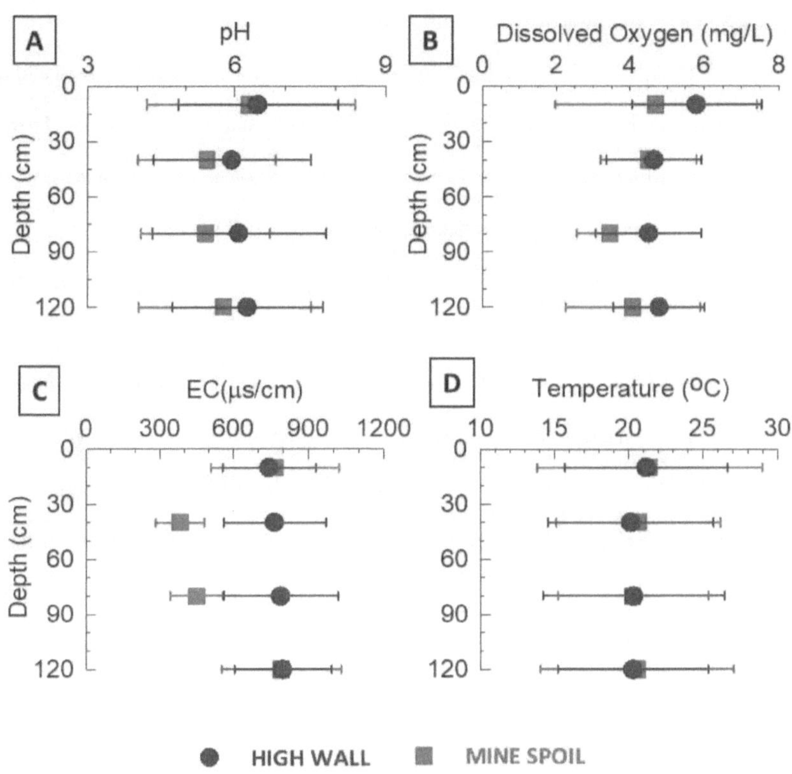

Figure 7: Average values for field measurements. pH(A), dissolved oxygen (B), electrical conductivity (C), and temperature (D) collected from lysimeters installed in High Wall soils (blue circles, left) and Mine Spoil soils (red squares, center) over the sampling season for lysimeters installed at 10 cm, 40 cm, 80 cm, and 120 cm. Singer et al., 2021.

4.2.5 Pore water aqueous metal concentrations

The first 10 cm of the HW had the highest concentrations of Fe and Al at the HW location, and the first 10 cm of the MS had the highest concentrations of Mn. There were no cations which were consistently higher in concentration all throughout the HW. Other trends which were in lower in concentrations include Cu found in its highest concentrations in the first 10 cm of the HW, and Ni found in its highest concentration in the first 10 cm of the MS.

At 5.00 ± 2.04 µmol/L Fe was $3.6 - 9$ times higher in the first 10 cm of the HW than it was for the rest of the profile, and it was 7.9 times higher than the first 10 cm of the MS. In the MS, Fe was $0.61 - 0.63$ µmol/L until it doubled in concentration when reaching 120 cm. The Fe profile for the HW was 5.00 ± 2.04 µmol/L at 10 cm, 1.39 ± 0.51 µmol/L at 40 cm, 0.55 ± 0.23 µmol/L at 80 cm, and 0.75 ± 0.56 µmol/L at 120 cm. The Fe profile for the MS was 0.63 ± 0.14 µmol/L at 10 cm, 0.61 ± 0.22 µmol/L at 40 cm, 0.63 ± 0.24 µmol/L at 80 cm, and 1.30 ± 0.75 µmol/L at 120 cm.

The HW had an even larger discrepancy with its shallow depth when looking at Al. At 49.90 ± 25.36 µmol/L the first 10 cm was 9.4 times higher than 40 cm depth, 26.8 times higher than 120 cm, and 14.5 times higher than the first 20 cm of the MS. The first 10 cm of the MS had only a half or third of the amount of Al that was found in the rest of the profile. The Al profile for the HW was 49.90 ± 25.36 µmol/L at 10 cm, 5.31 ± 1.83 µmol/L at 40 cm, 4.29 ± 2.21 µmol/L at 80 cm, and 1.86 ± 0.87 µmol/L at 120 cm. The

Al profile for the MS was 3.45 ± 1.90 μmol/L at 10 cm, 11.66 ± 5.52 μmol/L at 40 cm, 8.05 ± 4.11 μmol/L at 80 cm, and 9.72 ± 4.96 μmol/L at 120 cm.

Mn was higher at all depths in the MS than any of the HW samples. The 10 cm and 120 cm had the highest concentrations, with 5.95 ± 4.84 μmol/L and 4.25 ± 1.95 μmol/L, respectively. There was no trend with increasing or decreasing of Mn concentrations throughout the HW profile, with values that ranged between $0.76 - 1.69$ μmol/L. The Mn profile for the HW was 1.66 ± 0.66 μmol/L at 10 cm, 0.76 ± 0.98 μmol/L at 40 cm, 1.69 ± 1.22 μmol/L at 80 cm, and 1.01 ± 1.21 μmol/L at 120 cm. The Mn profile for the MS was 5.95 ± 4.84 μmol/L at 10 cm, 2.11 ± 0.85 μmol/L at 40 cm, 2.06 ± 1.27 μmol/L at 80 cm, and 4.25 ± 1.95 μmol/L at 120 cm.

Excluding the HW at 10 cm, Cu had the least amount of difference between the HW and MS at all depths. The HW and MS had Cu concentrations ranging from $0.15 - 0.18$ μmol/L at all depths except the first 10 cm of the HW, and at 120 cm for the MS. Cu was 0.56 ± 0.14 μmol/L at 10 cm of depth at the HW, which is 2.8 times higher than the 120 cm depth of the MS which was 0.20 ± 0.07 μmol/L. The Cu profile for the HW was 0.56 ± 0.14 μmol/L at 10 cm, 0.18 ± 0.03 μmol/L at 40 cm, 0.17 ± 0.03 μmol/L at 80 cm, and 0.15 ± 0.02 μmol/L at 120 cm. The Cu profile for the MS was 0.17 ± 0.03 μmol/L at 10 cm, 0.16 ± 0.02 μmol/L at 40 cm, 0.15 ± 0.03 μmol/L at 80 cm, and 0.20 ± 0.07 μmol/L at 120 cm.

Ni had an overall decreasing trend with depth in the MS, with it's high of 1.18 ± 0.71 μmol/L at 10 cm, and its low of 0.25 ± 0.13 μmol/L at 80 cm. Ni remained in higher concentrations in the MS than the HW, except for at 80 cm, where the HW was 92% higher. The Ni profile for the HW was 0.29 ± 0.04 μmol/L at 10 cm, 0.12 ± 0.08 μmol/L at 40 cm, 0.48 ± 0.22 μmol/L at 80 cm, and 0.17 ± 0.11 μmol/L at 120 cm. The Ni profile for the MS was 1.18 ± 0.71 μmol/L at 10 cm, 0.57 ± 0.25 μmol/L at 40 cm, 0.25 ± 0.13 μmol/L at 80 cm, and 0.39 ± 0.15 μmol/L at 120 cm.

Zn was found in higher concentrations at all depths in the MS profile, and Zn never exceeded a difference of 0.22 μmol/L when compared to its relative depth in the HW. The Zn profile for the HW was 0.74 ± 0.35 μmol/L at 10 cm, 0.48 ± 0.36 μmol/L at 40 cm, 0.59 ± 0.24 μmol/L at 80 cm, and 0.48 ± 0.30 μmol/L at 120 cm. The Zn profile for the MS was 0.88 ± 0.39 μmol/L at 10 cm, 0.70 ± 0.22 μmol/L at 40 cm, 0.78 ± 0.33 μmol/L at 80 cm, and 0.69 ± 0.24 μmol/L at 120 cm (Figure 8).

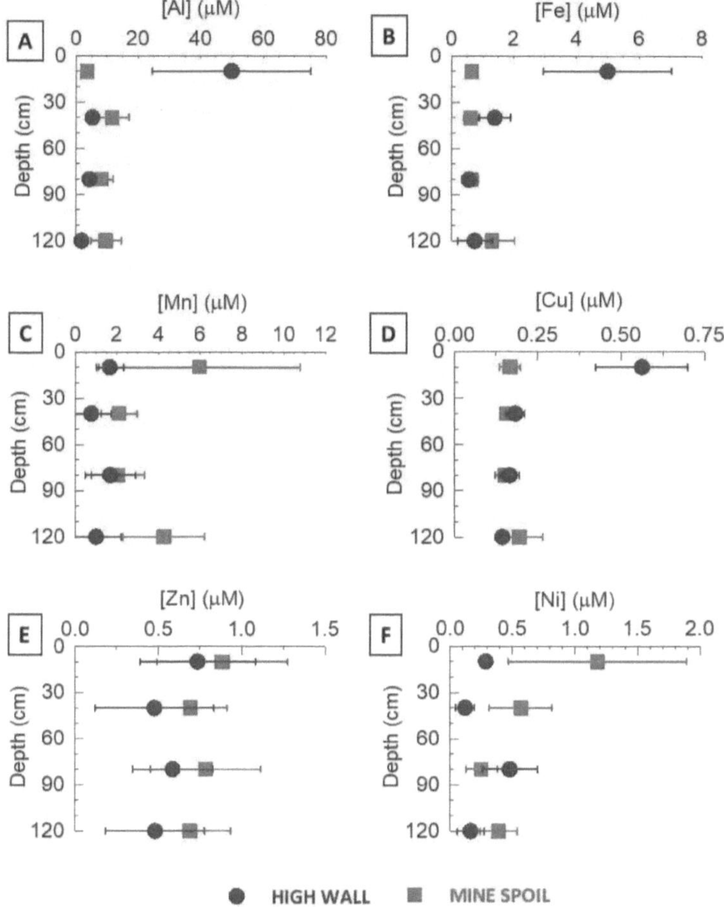

Figure 8: Average ICP-OES concentrations for cations in soil pore water. Al (A), Fe (B), Mn (C), Cu (D), Zn (E), and Ni (F) collected from lysimeters installed in High Wall soils (blue circles, left) and Mine Spoil soils (red squares, center) over the sampling season for lysimeters installed at 10 cm, 40 cm, 80 cm, and 120 cm. Singer et al., 2021.

4.2.6 Pore Water Anion Concentrations

Concentrations fluctuated greatly depending on the time of year the sample was taken (Tables 6.1 – 6.14). Concentration of all anions tested for the initial 10 cm, apart from NO_3^-, were always highest for MT location (there was no detectable Br^- in the HW for this depth). Average valves for lysimeters at 40, 80, and 120 cm of depth had higher concentrations of F^-, SO_4^{2-}, NO_2^-, in the MT location. Average values for lysimeters at 40, 80, and 120 cm had higher concentrations of Cl^-, Br^-, and PO_4^{2-} in the HW location. NO_3^- HW samples had higher concentrations at the 10 cm with a gradual decline with depth. The HW had increasing values of sulfate with depth, while the MS had its highest values in the 10 and 120 cm location.

The average concentration of fluoride in the HW is 55.73 ± 82.97 µmol at 10 cm, 13.96 ± 5.98 µmol at 40 cm, 58.95 ± 92.12 µmol at 80 cm, and 17.08 ± 5.22 µmol at 120 cm. The average concentration of fluoride in the MT is 128.93 ± 168.10 µmol at 10 cm, 60.32 ± 78.89 µmol at 40 cm, 93.94 ± 193.34 µmol at 80 cm, and 97.44 ± 155.93 µmol at 120 cm.

The average concentration of chloride in the HW is 234.52 ± 358.33 µmol at 10 cm, 401.98 ± 726.76 µmol at 40 cm, 229.72 ± 93.48 µmol at 80 cm, and 203.63 ± 220.37 µmol at 120 cm. The average concentration of chloride in the MS is 405.62 ± 464.00 µmol at 10 cm, 266.13 ± 377.15 µmol at 40 cm, 176.85 ± 270.29 µmol at 80 cm, and 191.70 ± 143.92 µmol at 120 cm (Figure 9).

The average concentration of nitrite in the HW is 86.65 ± 14.99 μmol at 10 cm, 64.85 ± 12.04 μmol at 40 cm, 64.61 ± 3.59 μmol at 80 cm, and 62.03 ± 2.97 μmol at 120 cm.

The average concentration of sulfate in the HW is 3496.77 ± 2515.20 μmol at 10 cm, 4724.29 ± 3392.02 μmol at 40 cm, 5600.00 ± 2880.85 μmol at 80 cm, and 5184.14 ± 2996.44 μmol at 120 cm. The average concentration of sulfate in the MS is 6176.85 ± 2452.92 μmol at 10 cm, 2273.15 ± 731.63 μmol at 40 cm, 2782.36 ± 1085.34 μmol at 80 cm, and 4214.70 ± 713.55 μmol at 120 cm (Figure 9).

The average concentration of bromide in the HW is 0 μmol at 10 cm, 412.93 ± 294.41 μmol at 40 cm, 416.28 μmol at 80 cm, and 411.21 μmol at 120 cm. The average concentration of bromide in the MS is 26.34 μmol at 10 cm, 155.33 ± 249.41 μmol at 40 cm, 20.79 ± 5.74 μmol at 80 cm, and 20.02 ± 9.06 μmol at 120 cm.

The average concentration of phosphate in the HW is 95.06 ± 144.14 μmol at 10 cm, 30.44 ± 14.38 μmol at 40 cm, 39.85 ± 2.70 μmol at 80 cm, and 90.08 ± 103.71 μmol at 120 cm. The average concentration of phosphate in the MS is 604.19 μmol at 10 cm, 60.30 ± 70.30 μmol at 40 cm, 35.46 ± 19.15 μmol at 80 cm, and 32.28 ± 13.65 μmol at 120 cm.

4.2.7 Pore Water Dissolved Organic Carbon (DOC):

The first 10 cm of the HW had 91.45 ± 23.61 mg/L of DOC; higher than any other sample depth in the HW or MS, and 5.38 times higher than the first 10 cm of the MS. The MS had little variance throughout its profile, with its highest concentration at 120 cm, with 27.08 ± 9.03 mg/L. Remaining average values for DOC include 14.16 ± 3.78 mg/L at 40 cm, 9.12 ± 2.75 mg/L at 80 cm, and 8.81 ± 2.28 mg/L at 120 cm at the HW, and 16.98 ± 6.42 mg/L at 10 cm, 11.43 ± 2.95 mg/L at 40 cm, 22.67 ± 5.67 mg/L at 80 cm at the MS (Figure 9).

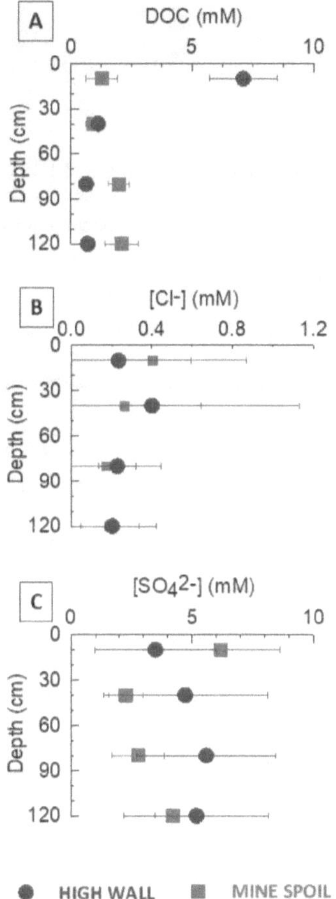

Figure 9: Lysimeter Data. Average concentrations for dissolved organic carbon (DOC) (A), chloride (B), and sulfate (C) collected from lysimeters installed in High Wall soils (blue circles, left) and Mine Spoil soils (red squares, center) over the sampling season for lysimeters installed at 10 cm, 40 cm, 80 cm, and 120 cm. Singer et al., 2021

35

4.2.8 Aluminum, iron, and manganese speciation:

The exchangeable Al (Figure 10, top panel) between the high wall and mine spoil core did not have any specific trend and was oftentimes higher than the concentrations found in the carbonate fraction; exchangeable Al comprised 1.1% of the HW profile, and 0.7% of the MS profile. The oxidizable fraction showed about double the concentrations of Al in the mine spoil than the high wall, comprising 0.5% of the profile for both profiles. The highest concentrations of Al were found in the reducible fraction comprising 75.2% of the HW profile, and 79.4% of the MS profile. The Al concentrations of the reducible fraction of the HW gradually decreased with depth, a trend that was not observed in the MS. Total extractable Al from the mine spoil core was 2,872 mmol kg^{-1}; total extractable Al from the HW was 2,121 mmol kg^{-1}.

Iron concentrations were highest in the reducible fraction (comprised about 98.6% of the Fe in both the HW and MS) and did not show significant changes with depth with the carbonate (<0.02%) and oxidizable fraction (1.3%) (Figure 10, middle panel). Total extractable Fe from the mine spoil core was 10,360 mmol kg^{-1}; total extractable Fe from the HW was 7,281 mmol kg^{-1}. Overall, Fe levels could be found in higher concentrations between 50-100 cm depth.

Mn had little variance throughout the HW profile (Figure 10, bottom panel), with slight enrichment in its surface layers. Exchangeable Mn in the HW was only detected in the first 10 cm and comprised only 1% of total extractable Mn for its profile. The HW carbonate fraction was also more concentrated throughout the first 10 cm of depth, with the total profile comprising < 3% of total extractable Mn. The remaining 96% of

extractable Mn in the HW was from the reducible fraction. This portion was enriched in

the first 30 cm of the profile, and again between 90-110 cm depth. Extractable Mn in the

MS was significantly higher than in the HW, totaling 272 mmol kg^{-1} in contrast to the

HW's sum of 21 mmol kg^{-1}. Unlike the HW extractions, Mn was found throughout all

depths during the exchangeable fraction, with enrichment in the first 10 cm of the profile.

Although Mn bound to carbonates in the MS profile only comprised 0.6% of its total

extractable Mn concentrations, its presence was more than double what was found in the

carbonate fraction of the HW. 97% of extractable Mn of the MS was found in the

reducible fraction, with enrichment found within 70 – 100 cm of depth. There was no

detectable oxidizable Mn for either profile.

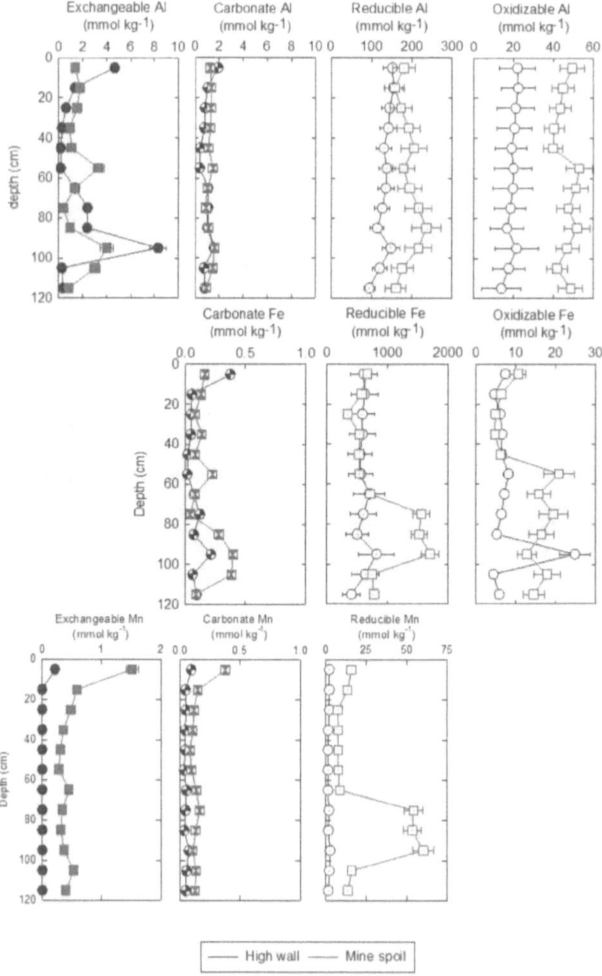

Figure 10: Average concentrations of Al, Mn, and Fe during sequential extraction procedure. There was no detectable exchangeable Fe or oxidizable Mn in either soil column. Singer et al., 2021

5. DISCUSSION

5.1 Mineral Weathering

Samples from the HW and MS were similar in mineralogy and particle size. With visual evaluation of the soil core the HW had a more developed O and A soil horizon providing rich plant-derived organics to the profile, and there was more coal content in the MS especially when reaching 50 cm depth providing rock-derived organic matter on the lower half of the profile. Using the USDA Soil Texture Triangle, all of the HW and MS samples would be classified as a sandy loam, with exception of HW 5 cm, 15 cm, and 115 cm samples, which are classified as a silt loam. Overall, the mine spoil had slightly more clay at all depths when compared to the high wall (average difference of 1.9% more clay in the MS) (Figure 11).

XRD analysis showed the bulk composition was dominated by quartz and muscovite. Other minerals present included feldspar, kaolinite, and chlorite. The presence of kaolinite (formed through the complete breakdown of feldspar or mica) from XRD analysis, and Al concentrations from sequential extraction indicates a moderate weathering profile (Strawn et al, 2015). The absence of pyrite, and the presence of kaolinite and chlorite indicate that primary weathering has taken place. Pyrite dissolution will drop the pH of the environment, but during our sampling period the pH of the MS location was not dramatically lower than the HW location. With evidence of pyrite dissolution found throughout the MS soil, there must have been mineral buffers present. When muscovite and chlorite weather they can neutralize acid though soluble ion release (Clark et al., 2018). Calcium carbonate found in limestone would be able to buffer the pH during pyrite dissolution. We did not incorporate the study of calcium into our cation study, but it has likely played a key factor into our pH levels due to the lithology of our location, and through the finding of shell fossils during core excavation (Figure 11).

The first 10 cm of soil would have the most exposure to oxygen, allowing for pyrite dissolution. Pyrite dissolution leads to sulfate elution, and the first 10 cm of the MS had the highest concentrations of sulfate when compared to any other sampling location (Clark et al., 2018). The HW had sulfate trends of increasing concentration with depth, while the MS had very high concentration at 10 cm then drops and starts rising at depth but is less than HW at all values except 10 cm.

Aqueous Mn and Ni are also found in high concentrations in the first 10 cm of the MS. Clark, et al. (2018) found a positive correlation to muscovite and chlorite point counts and peak Mn and Ni release. Clark concluded that chlorite, a secondary mineral derived from muscovite, can be highly reactive in mined settings, and can contribute to releasing soluble Mn and Ni from spoil (Clark et al., 2018). Although the presence of these trace minerals may contribute to metal release, our study points to organic matter as the main influence on whether metals become labile. The presence of kaolinite and chlorite are indicative of intermediate weathering, and their abundance has influence over adsorption of metals to clay within the profile.

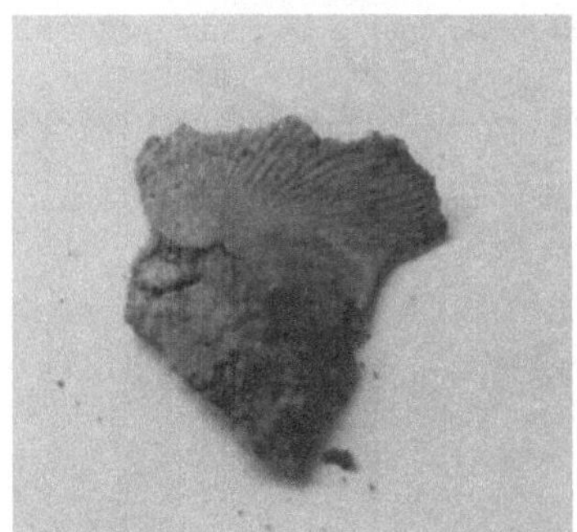

Figure 11: Shell fossil found during excavation
of soil cores. Photo by Laura Zemanek

EC readings from the high wall had little variability with depth (±54.93 µS/cm), while the MS readings showed lower levels at 40 and 80 cm, 381.60 µS/cm and 450.53 µS/cm, respectively. A study by Mukhopadhyay et al. (2016) showed that tree root exudates can influence the solubility of soil compounds, and there were also changes in EC with different weathering profiles. Differences between the HW and MS hill that should be considered are vegetation and particle size. With a more robust vegetative cover on the HW, evapotranspiration can concentrate the solution, making EC readings higher in its profile. The two lowest EC readings were located in the MS 40 and 80 cm lysimeters. These were located at depths where clay content was high, resulting in a lower EC due to slower weathering rates and less ion exchange due to the stability of clay.

Fe, Al, and Mn were analyzed during the sequential extraction procedure. All three metals were released in highest concentrations during extraction of the reducible fraction indicating the presence of (oxy)hydroxides, and all metals were found in higher concentrations in the MS profile. Fe was below the detection limit for both cores for the first solution (exchangeables).

There was significantly more extractable Al in the MS profile (2872.12 mmol kg^{-1}) vs. the HW profile (212.27 mmol kg^{-1} total extractable), with the majority being concentrated in the reducible portion but also about a quarter of its total concentration released during the oxidizable extractions.

Fe and Mn were affected most by the rock-derived organic matter in the MS (from 70 – 120 cm), becoming more labile. These elements are more likely to be sequestered to weathered products from the coal layers, and therefore were not found as abundant cations in soil pore water. There was no Mn detectable with the oxidizable fraction, and with exception of HW 10, there was no detectable Mn within the HW exchangeable fraction. The presence of pyrite in the MS highly affected the overall content of Mn found during the extractions, with having a total extractable content of 272.33 mmol kg^{-1} in the MS and only 21.62 mmol kg^{-1} in the HW. Formation of secondary minerals with the incorporation of Fe and Mn as we have in Huff Run can lead to long term fluxes in metal release over time.

5.2 Trace Metal Transport

Soil organic matter (SOM) was tested for through LOI and DOC. Even though the HW organic carbon was lower, as determined by LOI testing, there was significantly more DOC in the first 10 cm of the HW, which then drastically decreased and then fell below the level of the MS at 80 cm. The HW has had more time to develop its soil (the MS may have only had 50 years to develop), which includes plant, animal, and microbial activity, which was effects were also observed in color when taking samples. With its stable LOI content and greater concentration of DOC in the upper portion of the profile, the largest contributor is biologic activity. The highest DOC was located in the plant-derived organic rich layer.   SOM is highly reactive and can contain ligands which increase the solubility of metals (Strawn et al., 2015). This portion of the profile is also where we had the highest concentrations of Al, Fe, Cu in the pore water.

In contrast, the MS had a variable LOI, with its highest concentrations between 50 – 80 cm. While extracting the soil core we found this area to be a darker gray color, while the HW was brown throughout. There were also more soil aggregates and black nodules in this area. Soil aggregation rate shows a strong correlation to organic carbon in soil, and can be sequestered in macroaggregates (Li et al., 2016). Unlike the high DOC found in the HW where organics are rich, there was not a high DOC found in the organic rich layers of the MS. Our results show that there may be a correlation with the lability of carbon being high when in areas of plant-derived organics, and low in areas of rock-derived organics.

In order for H^+ to be toxic to plants, it must be less than a pH of 3, which we did not find at either site within our testing period. When comparing the average pH values for the field season, the MS had more acidic values at all depths (Appendix 12). Isolating and comparing the ICP-OES data for Fe and Al concentrations showed no correlation to the abundance of either element in being a predictor to pH. Pyrite oxidation can lower pH values to less than 2, and once the pyrite minerals are exhausted, the pH will start to rise as the acidity is leached out and alkaline minerals dissolve (Strawn et al., 2015). The pH values of the MS and HW are slightly acidic, indicating that pyrite dissolution has completed (or near completed), or is being actively buffered. During XRD analysis pyrite was not detected, but a study from Singer et al. (2021) used scanning electron microscopy (SEM) for this sampling location and found pyrite coated in either Fe-O-bearing or Si-Al-O-bearing phases.

A study by Monterroso et al. (1994) looked at aluminum and iron speciation among mine spoil hills of different ages. They found that older areas had a lower pH, higher Eh, and high concentrations of Fe and SO_4^{2-} and Al. The speciation of Fe and Al complexes at pH ranges of $5.0 - 6.0$ differed from those at pH greater than 6.0. MS average pH values for 40, 80, and 120 cm fall within the $5.0 - 6.0$ range, making them likely to contain Al-F monomers and $Fe(OH)_2^+$ complexes. All other locations we sampled are likely to have Al-OH and $Fe(OH)_3$ complexes. According to Monterroso et al. (1994) all of the aluminum in our profiles are likely labile, the Fe is present as Fe^{2+}, and if pyrite is still available the pH will decrease with time if reclamation is not performed.

Fe pore water trends showed that the first 10 cm of the HW was very high (5 μmol/L), then drops and is uniform, MS Fe was low (0.63 μmol/L) and uniform throughout. The high wall had ~8 times more Fe in the first 10 cm than the MS, indicating that plant-derived organic material may be causing high mobility of iron in the profile. Although the Fe was lower in the pore water, the sequential extraction results show that the Fe content is higher in the MS profile, indicating that the Fe is bound into what is likely iron oxyhydroxides. The ion with the greatest abundance in the soil pore water was the Al in the first 10 cm of the HW profile, with a concentration of 49.9 μmol/L, which is most likely a result of high plant-derived organic matter being present in that layer. Another element that seems to have mobility highly effected by plant-derived dissolved organic matter is Cu. The first 10 cm of the HW has over 3 times the concentration of Cu, with 0.56 μmol/L. The rest of the profile was only greater in the HW by $0.02 - 0.05$ μmol/L, indicating that plant organics is the most significant factor for its release from minerals.

Mn was in higher concentrations in the MS. Mn can become soluble in the presence of pyrite dissolution and can also be released during pyrite dissolution if it was incorporated into the structure by substitution (Huerta-Diaz & Morse, 1991). Singer et al. (2021) found that the majority of solid phase Mn from our profile was associated with metal (oxy)hydroxides in the reducible fraction. He also found that elements such as As and Se are being sequestered from our profile by these Fe- and Mn-bearing phases (Singer et al., 2021). Having similar lithology yet finding more soluble Mn in the MS is further support for pyrite dissolution. When soluble Mn is in acidic and circumneutral pH, it is stable, which leaves it in higher concentration in MS settings (Herndon et al., 2019).

Zn had similar trends found in the HW and MS, with 34% more Zn found in the MS. A study by Cancés et al. (2003) looked at dissolved organic matter and its effects on metal mobility. Cancés found that while Cu is more labile when exposed to DOC, Zn is typically in solution as free aquo-ions. Ni was found in slightly higher concentrations in the MS, and slightly decreased in concentrations throughout the field season.

The DO of soil pore water from the high wall was consistently higher than the mine spoil (Appendix 13). The lower MS DO could be a result of pyrite oxidation coupled to oxygen reduction reactions which will release Fe and S that bind with the oxygen to form other compounds. Overall, the MS depths of 40 and 80 cm follow the same trend as the HW, but the amounts they increased or decreased varied. DO in the HW was very high in first 10 cm, then drops and is uniformly low. The first 10 cm of the HW was expected to have the highest DO as it is exposed to surface conditions and has more biologic activity (such as more extensive root systems and potentially more extensive earthworm colonies), and less reduction reactions taking place due to the lack of pyrite dissolution reactions. Over the course of the field season the DO for both sites decreased, most likely to oxygen solubility decreasing as the temperature increased.

Barfoud et al. (2019) used steady flow of water with different concentration of dissolved oxygen to circulate through mine tailings which had varying amounts of pyrite content. Their study included sulfate leachate that varied between 65 – 140 mg/L. The sulfate leachate collected from our study had most samples well above 140 mg/L, with some readings above 1000 mg/L. The HW overall had higher readings of sulfate concentrations. Barfoud found that conductivity, sulfate concentrations, and release of metals into solution were strongly correlated to dissolved oxygen. Our MS hill has gone through its primary stage of weathering where most of the pyrite dissolution has already taken place.

Dissolved metal concentrations from the HW are highest in the first 10 cm, and then decrease with depth which aligns with the HW's dissolved oxygen content influencing metal release. With DO available to the residual coal layers, the pH was lower than other areas of the profile, but not often less than 5, indicating that oxidative weathering is low. In the MS, Mn, Ni, and Zn were in its highest concentrations in the first 10 cm of the profile, but the other metals had little variance with depth, with slight increase towards 120 cm. Cravotta (1993), had similar findings when studying ground water running through reclaimed mine spoil. He discovered that pyrite oxidation in the unsaturated zone does not rely on oxygen alone but can also oxidize when exposed to ferric iron. The interaction between pyrite and ferric iron can be a contributor to why DO was inversely related to sulphate concentrations in the pore water when comparing the HW and MS.

Low pH and high concentrations of aqueous sulfate indicate pyrite oxidative

dissolution reactions which readily occur with percolating water and during the

weathering process (Strawn et al., 2015). Throughout the duration of the sampling

season SO_4^{2-} decreased. This may be directly related to the lower levels of dissolve

oxygen present as the field season progressed (Barfoud, 2019).

6.   CONCLUSION

Non-point sources of coal AMD formed through mine spoil shows differences in

mineral and pore water analysis. Fluctuating weather conditions, weathering rates, and

altering conditions which influence materials to switch back and forth between reducing

and oxidizing environments can produce swings in soil and water pH and cause fluxes in

metal loading of streams. Drops in pH and rises in metal concentrations lead to losses of

life and diversity in the plant and wildlife population of a watershed.

In 2010, The Huff Run Watershed flagged HR-25 as contributing to 4% of iron

spikes to two mainstreams in the watershed during low and medium flow conditions, and

5% during high flow (ODNR, 2013). The watershed project recognized non-point source

contamination treatment as imperative in order to recover multiple areas of the Huff Run

Watershed from the damage of heavy metal leaching from AMD hills. The JS&L AMD

Reclamation Project was completed October 16, 2015, installing limestone channels, limestone leach bed and precipitation basin (NPS, 2018).

In 2010-2015, all 16 km of the Huff Run Watershed met its minimum stream pH goal of 6.5, but in 2016 the stream 3.2 km (impacted by HR-25) dropped to 6.4 and did not meet its goal. The 2017-2018 report indicated that once again all 16 km of the Huff Run Watershed has met its minimum stream pH goal of 6.5. An NPS map of biological quality for Huff Run from the 2017-2018 report shows that the area receiving water from HR-25 has macroinveterbrate aggregated between 0 – 7, which indicated "very poor" diversity and abundance of macroinvertebrates, while upstream in the watershed there are ratings of either "poor" or "good" (MAIS, 2). There is no updated ratings, as yearly monitoring is not common due to lack of resources (MAIS, 2).

Our study sampled one native hill, and one mine spoil hill, taking soil cores to 120 cm of depth at each location. The samples taken for this book project were from May 20, 2015 – October 2, 2015. The overall composition of the mine spoil hill is very similar to the native hill, as it is taken from the native hill, but the mine spoil has higher amounts of clay minerals (shown through XRD analysis) which are also available for reactions, adsorption, and desorption. LOI data results have an almost uniform appearance for the native hill, but the mine spoil hill has increases in any area with high amounts of discarded coal.

Sequential extraction found high concentrations of iron, manganese, and aluminum in the reducible fraction that can become leached into the pore water when reducing conditions are met. Although XRD analysis did not find detectable levels of pyrite, a study performed by Singer et al. (2021) performed SEM analysis on the samples used in this book project. He found that there is residual pyrite coated in metal (oxy)hydroxides, along with Fe- and Mn- bearing species in reducible form. The magnitude of metal leaching that can occur when all the mine spoil hills are considered can be devastating to plant and wildlife in the watershed. As more vegetation is established on the mine spoil broadening the O and A horizons there is also more potential of reactions between organic matter and sulfides with metals such as Cu and Zn (Strawn et al., 2015).

Overall, metals in soil pore water were in their highest concentrations within the first 10 cm, and decreased with depth, while those bound to minerals were fairly uniform with increases for the MS around 50 cm depth. Carbon and metal mobility was influenced differently by the presence of plant-derived vs. rock-derived organics, DO, particle size, and aggregate formation. Where levels of DOC were high, there were higher concentrations of soluble Al, Fe, and Cu. In areas of heavy clay and residual coal (supported through LOI and PSA data), Fe, Mn, and Al were found bound to minerals as found through the reducible fraction of our sequential extraction procedure.

The weathering profile is moderate, and has already gone through primary weathering, which has resulted in both hills having a similar pH, but leaves the potential for metal leaching fluxes if put back into reducing conditions. The increase of metal concentration of the MS samples from 50 – 120 cm released during sequential extraction is likely due to metals released from sulfides which can adsorb to iron oxydyroxides from the secondary phase of weathering. We found it important to include datasets that went to depths of at least 120 cm, as much of the current available research of non-point sources has more shallow testing depths (commonly up to 30 cm depth). Further research recommendations for Huff Run Watershed would be to include more sampling locations, testing over a longer period of time, and collecting data for Eh and calcium concentrations. As shown through Huff Run Watershed data one area, such as HR-25, can influence several kilometers running downstream of the contamination.

References

ATSDR, 2007 "Division of toxicology and environmental medicine ToxFAQs"

ARSENIC #7440-38-2

Babcock, L.E., 2009 "Visualizing Earth History" John Wiley & Sons, Inc.

Baird, G.C., and C.W. Shabica, 1980, The Mazon Creek depositional event; examination

of Francis Creek and analogous facies in the Midcontinent region: in Middle and

Late Pennsylvanian Strata on Margin of Illinois Basin, Vermilion County, Illinois,

Vermillion and Parke Counties, Indiana (R.L. Langenheim, editor). Annual Field

Conference—Society of Economic Paleontologists and Mineralogists. Great lakes

Section, No. 10, p. 79–92.

Baker, J., 2015 "Visitors get close look at Huff Run rehabilitation plan"

TimesReporter.com Accessed 18 Nov. 2015

Barfoud, L., Pabst, T., Zagury, G.J., Plante, B., 2019 "Effect of Dissolved Oxygen on

the Oxidation of Saturated Mine Tailings" Geo-Environmental Engineering, p.5

Berner, R.A., 1970 "Sedimentary Pyrite Formation" American Journal of Science, vol.

268, pp.1-23

Bowman, J., Johnson, K., 2012. "2012 Stream Health Report: An Evaluation of

Water Quality, Biology, and Acid Mine Drainage Reclamation in Five

Watersheds: Raccoon Creek, Monday Creek, Sunday Creek, Huff Run, and

Leading Creek" *Voinovich School of Leadership and Public Affairs at Ohio*

University.

Brownmag, O., Brown, M., 2018 "Effects of Soil Temperature on Some Soil Properties

    and Plant Growth" MedCrave Advances in Plants & Agriculture Research, vol. 8

    Issue 1, pp. 34 – 37

Cancés et al., 2003 "Metal ions speciation in soil and its solution: experimental data and

    model results" Geoderma 113, pp. 341-355

Crowell, D.L., 2005, "History of Coal Mining in Ohio" *Ohio Department of Natural*

    Resources. GeoFacts No. 14. Accessed 6 Feb. 2015.

    http://www.ohiocoal.com/downloads/history-ohio-coal-mining.pdf

Cravotta, C.A. III 1993 "Secondary Iron-Sulfate Mineral as Sources of Sulfate and

    Acidity: Geochemical Evolution of Acidic Ground Water at a Reclaimed Surface

    Coal Mine in Pennsylvania" Environmental Geochemistry of Sulfide Oxidation,

    Chapter 23, pp. 345 - 364

Devasahayam, S. 2006 "Application of Particle Size Distribution Analysis in Evaluating

    the Weathering of Coal Mine Rejects and Tailings" Fuel Processing Technology

    88 (2007) 295 – 301

Divvela, P. 2010 "Sequential Extraction Procedure" Lab Science News – Technical Blog.

    Accessed 9 March 2015

    http://www.caslab.com/News/sequential-extraction.html

EPA, 2005 "2005 Targeted watershed grants: Huff Run Ohio" *Targeted* Watersheds,

    EPA840-F-07-001C

EPA, 2011 "Point Source, 40 CFR 122.2" Accessed 28 January 2015.

http://www.epa.gov/region6/6en/w/ptsource.htm

Feldman, C. 1983 "Behavior of Trace Refractory Minerals in the Lithium Metaborate Fusion-Acid Dissolution Procedure" Anal. Chem., vol 55, no. 14, 2451-2453

Fetter, C.W. 2001 "Applied Hydrogeology" 4 ed., Upper Saddle River, NJ: Prentice Hall, pp.85   Table 3.7

Fleming, G., 2010. "Huff Run Watershed Acid Mine Draingae Abatement and Treatment Plan" Ohio Department of Natural Resources

Friedman, S.P., 2005 "Soil Properties Influencing Apparent Electrical Conductivity: A Review" The Institute of Soil, Water and Environmental Sciences, Agricultural Research Organization, The Volcani Center, Bet Dagan 50250, Isreal. Computers and Electronics in Agriculture 46, pp.45 – 70.

Gerber, T.D., Buzard, R.W., 1983. "Soil Survey of Carroll County, Ohio. United States Department of Agriculture"

Herndon et al., 2019 "Iron and Manganese Biochemistry in Forested Coal Mine Spoil" Soil Systems 3, 13; doi:10.3390/soilsystems3010013, pp. 1-19

Hoffert, R.J., 1947 "Industrial Wastes…Acid Mine Drainage", *Industrial and Engineering* Chemistry, Pennsylvania Department of Health, Harrisburg PA.

Huerta-Diaz M.A., Morse J.W., 1991 "Pyritization of trace metals in anoxic marine sediments" Geochimica et Cosmochimica Acta Vol. 56, pp. 2681-2702

Ingamells, C.O., 1970 "Lithium Metaborate Flux in Silicate Analysis" Anal., Chim. Acta, 52, 323-334

Jenny, H., 1994 "Factors of Soil Formation: A System of Quantitative Pedology" *Dover Publications, Inc. New York.* Forward by Ronald Amundson

Kinney, C., 2013. "Huff Run Acid Mine Drainage Abatement and Treatment Plan Addendum". Ohio Department of Natural Resources Division of Mineral Resources Management New Philidelphia Ohio

Kleinmann, R., 1989. "Acid Mine Drainage: US Bureau of Mines Researches Control Methods for Coal and Metal Mines" *US Bureau of Mines*

Larsen, D., Mann, R., 2005 "Origin of High Manganese Concentraitons in Coal Mine Drainage, Eastern Tennessee" *Journal of Geochemical Engineering,* pp.143-163

LeHigh University (2000 – 20011) "Basic AMD Chemistry", Enviro Sci Inquiry. 2011. Accessed 28 January, 2015.

http://www.ei.lehigh.edu/envirosci/enviroissue/amd/links/science2.html

Li et al., 2016 "Association of Soil Aggregation with the Distribution and Quality of Organic Carbon in Soil along an Elevation Gradient on Wuyi Mountain in China" Plos One 11(3): e0150898. Doi:10.1371/journal.pone.0150898

Lottermoser, Bernd, G. 2010 "Mine Wastes: Characterization, treatment and environmental impacts" 3rd edition. Springer. DOI 10.1007/978-3-642-12419-8

McLean, J.E., Bledso, B.E., 1992, "Behavior of Metals in Soils" *Ground Water Issue EPA, Office of Research and Development, Office of Solid Waste and Emergency Response.* EPA/540/S-92/018.

Medlin, J.H., Suhr, N.H., Bodkin, J.B., 1969 "Atomic Absorption Analysis of Silicates

Employing LiBO$_2$ Fusion" Atomic Absorption Newsletter, vol. 8, no. 2, 25-29

Monterroso C., et al., 1994 "Speciation and solubility control of Al and Fe in minespoil

solutions" The Science of the Total Environment 154, pp. 31 – 43

Mukhopadhyay S., et al., 2016 "Soil Quality Index for Evaluation of Reclaimed Coal

Mine Spoil" Science of the Total Environment 542, pp. 540 – 550

Nalbantov, Georgi I. et al. 2010 "Image Mining for Intelligent Autonomous Coal

Mining." *ResearchGate*, p.3

NPS, 2014 "Huff Run Watershed Report" Huff Run Watershed Partnership Inc., pp. 68.

Accessed 18 Nov. 2015

http://www.watersheddata.com/userview_file.aspx?UserFileLo=1&UserFileID=1

77

NPS, 2017-2018 "Collection of AMD Projects Report", p. 127

Non-Point Source Monitoring System, 2013 "Timeline of the Huff Run Watershed

Project Milestones & AMD Projects" 2013 NPS Report , Huff Run Watershed

ODNR Division of Mineral Resources, 2016 "History of Coal Mining & Regulation in

Ohio" Accessed 27 Nov. 2016. http://minerals.ohiodnr.gov/citizen-

resources/history-of-ohio-coal-mining

ODNR, 2005 "History of coal mining in Ohio" Division of Geology Survey, Geofacts

No. 14 Accessed 12 Nov. 2015.

http://minerals.ohiodnr.gov/portals/minerals/pdf/coal/geof14.pdf

ODNR, 2010 "Citizens Guide to Mining and Reclamation in Ohio" 2011.

Ohio Department of Natural Resources Division of Mineral Resources Management. Mining History. Accessed 6 Feb. 2015. http://minerals.ohiodnr.gov/portals/minerals/pdf/coal/permitting/citizens_guide.pdf

ODNR, 2011 "Citizens Guide to Mining and Reclamation in Ohio" Ohio Department of Natural Resources Division of Mineral Resources Management. Mining History. Accessed 18 Nov. 2015.

Ohio Division of Geological Survey, 2006, Bedrock geologic map of Ohio: Ohio Department of Natural Resources, Division of Geological Survey Map BG-1, generalized page-size version with text, 2 p., scale 1:2,000,000

Ohio Coal Association, 2015 "Ohio coal quick facts" Accessed 12 Nov. 2015. http://www.ohiocoal.com/learning-center/quick-facts.php

Ohio Coal Association "History of Ohio Coal Mining" 2014. Accessed 13 March 2015. http://www.ohiocoal.com/learning-center/history-coal-mining.php

Ohio Environmental Protection Agency, 2005 "2005 Targeted Watershed Grants: Huff Run Ohio" Water.epa.gov/grants_funding/twg/upload/2007_04_04_watershed_initiative_2005_huff_run.Pdf

Ohio University, 2015 "Huff Run Watershed Restoration Partnership" Ohio Watershed Data.   Accessed 27 Nov. 2016. http://www.watersheddata.com/huff.aspx

Perfrial, N., Rivera, N., Thompson, A., O'Day, P.A., Chorover, J., 2011. Trace contaminant concentration affects mineral transformation and pollutant fate in hydroxide-weathered Hanford sediments. Journal of Hazardous Materials, 197: 119-127

Pond, G.J. 2004, "Effects of Surface Mining and Residential Land Use on Headwater Stream Biotic Integrity in the Eastern Kentucky Coal Field Region". Kentucky Department of Environmental Protection, Division of Water, Water Quality Branch, Frankfort, KY.

Pratt, S.E., 2014 "Natural arsenic levels in Ohio soils exceed regulatory standards" *American Geosciences Institute*

Reece, B.A., 1995, "Acid Mine Drainage: Perpetual Pollution" *Mineral Policy Center*

Rocha-Nicoleite, E., Overbeck, G.E., Muller, S.C. 2017 "Degradation by coal mining should be priority in restoration planning" *ResearchGate* p. 2

Singer et al., 2021 "Biogeochemical Controls on the Potential for Long-Term Contaminant Leaching from Soils Developing on Historic Coal Mine Spoil" *Soil Systems* 5,3. https://doi.org/10.3390/soilsystems5010003

Strawn, D.G., Bohn, H.L., O'Connor, G.A., 2015, "Soil Chemistry" Wiley Blackwell. Ed.4 pp. 64, 328

Swaine, 1955 "Selenium in Nutrition" *National Research Council (US) Subcommittee on Selenium*. Washington (DC) National Academies Press (US)

Suhr, N.H. and Ingamells, C.O., 1996 "Solution Technique for Analysis of Silicates" Anal. Chem., 38, 730-734

Tessier, A., et al. 1979 "Sequential Extraction Procedure for the Speciation of Particulate

Trace Metals" Analyitcal Chemistry , Vol. 51, NO. 7, June 1979.

University of California Museum of Paleontology, 2009 "The Carboniferous Peiod"

http://www.ucmp.berkeley.edu/carboniferous/carboniferous.php. Accessed 5 Aug

2017

USDA 2001 "Table 1: Example of a Minimum Data Set of Indicators for Soil Quality"

Guidelines for Guidelines for Soil Quality Assessment in Conservation Planning

in Conservation Planning Conservation Planning. *United States Department of*

Agriculture, Natural Resources Conservation Service Soil Quality Institute. Pp.7

http://www.nrcs.usda.gov/Internet/FSE_DOCUMENTS/nrcs142p2_051259.pdf

U.S. EPA. 1994 "Method 200.7: Determination of Metals and Trace Elements in Water

and Wastes by Inductively Coupled Plasma-Atomic Emission Spectrometry"

Revision 4.4 Cincinnati, OH

U.S. Energy Information Administraiton, 2016 "Coal Explained: Where our coal comes

from" Accessed 27 Nov. 2016.

https://www.eia.gov/energyexplained/index.cfm?page=coal_where

Voinovich School of Leadership and Public Affairs at Ohio University

USGS, 2006 "USGS mine drainage activities" USGS Mining-Environment Studies,

accessed 12 Nov. 2015. http://mine-drainage.usgs.gov/sum/.

Waters,D.D., Roth, L., 1986. "Soil survey of Tuscarawas County, Ohio" *United States*

Department of Agriculture.

Willman, H.B., and J.N. Payne, 1942, Geology and Mineral Resources of the Marseilles,

Ottawa, and Streator Quadrangles: Illinois State Geological Survey Bulletin 66,

388 p.

Wise, M., 2005. "Huff Run Watershed Plan" *Huff Run Watershed Restoration

Partnership, Inc., Mineral City, Ohio*

APPENDICES

1. Web Soil Survey

Tuscarawas County, Ohio (OH157)

Map Unit Symbol	Map Unit Name	Acres in AOI	Percent of AOI
BkC	Berks shaly silt loam, 8 to 15 percent slopes	45.3	0.60%
BkD	Berks shaly silt loam, 15 to 25 percent slopes	34.2	0.40%
BkE	Berks channery silt loam, 25 to 35 percent slopes	2	0.00%
BnC	Bethesda channery clay loam, 8 to 15 percent slopes	22.1	0.30%
BnF	Bethesda channery clay loam, 25 to 70 percent slopes	687	8.70%
BtB	Bogart variant loam, 3 to 8 percent slopes	10.9	0.10%
CkB	Chili gravelly loam, 3 to 8 percent slopes	4.7	0.10%
CpB	Coshocton silt loam, 3 to 8 percent slopes	24.3	0.30%
CsC	Coshocton-Guernsey silt loams, 8 to 15 percent slopes	22.1	0.30%
CsD	Coshocton-Guernsey silt loams, 15 to 25 percent slopes	284.7	3.60%
FcA	Fitchville silt loam, 0 to 3 percent slopes	105.5	1.30%
FcB	Fitchville silt loam, 3 to 8 percent slopes	33.4	0.40%
FeB	Fitchville-Urban land complex, undulating	25.6	0.30%
GfB	Glenford silt loam, 3 to 8 percent slopes	16.4	0.20%
GfC	Glenford silt loam, 8 to 15 percent slopes	292.2	3.70%
HeC	Hazleton channery loam, 8 to 15 percent slopes	88.3	1.10%
HeD	Hazleton channery loam, 15 to 25 percent slopes	50.7	0.60%
HeE	Hazleton channery loam, 25 to 40 percent slopes	803.6	10.10%
HeF	Hazleton channery loam, 40 to 60 percent slopes	17.1	0.20%
HgF	Hazleton extremely bouldery loam, 25 to 60 percent slopes	74.4	0.90%
KeC	Keene silt loam, 8 to 15 percent slopes	46.2	0.60%
Or	Orrville silt loam, occasionally flooded	190.3	2.40%
RuA	Rush silt loam, 0 to 3 percent slopes	31.6	0.40%
Ua	Udorthents, hilly	7.3	0.10%
WbA	Weinbach silt loam, 0 to 3 percent slopes	33.3	0.40%
WhC	Westmoreland silt loam, 8 to 15 percent slopes	18.4	0.20%
WhD	Westmoreland silt loam, 15 to 25 percent slopes	59.2	0.70%
WhE	Westmoreland silt loam, 25 to 35 percent slopes	8.5	0.10%
WnC	Westmoreland-Guernsey silt loams, 8 to 15 percent slopes	12.5	0.20%
Subtotals for Soil Survey Area		**3,051.90**	**38.50%**
Totals for Area of Interest		**7,930.20**	**100.00%**

1.1 Web Soil Survey for Tuscarawas County, Ohio

Carroll County, Ohio (OH019)

Map Unit Symbol	Map Unit Name	Acres in AOI	Percent of AOI
BkC	Berks channery silt loam, 8 to 15 percent slopes	17.2	0.20%
BkD	Berks shaly silt loam, 15 to 25 percent slopes	60.9	0.80%
BkE	Berks channery silt loam, 25 to 35 percent slopes	11.9	0.10%
BnD	Bethesda channery clay loam, 8 to 25 percent slopes	105.2	1.30%
BnF	Bethesda channery clay loam, 25 to 70 percent slopes	663.7	8.40%
CnB	Coshocton silt loam, 3 to 8 percent slopes	62.5	0.80%
CoB	Coshocton-Keene silt loams, 3 to 8 percent slopes	168.5	2.10%
CuB	Culleoka silt loam, 3 to 8 percent slopes	322.6	4.10%
EbB	Elba silty clay loam, 3 to 8 percent slopes	2.8	0.00%
Ek	Elkinsville silt loam, rarely flooded	5.5	0.10%
FaD	Fairpoint channery clay loam, 8 to 25 percent slopes	99.7	1.30%
FcB	Fitchville silt loam, 3 to 8 percent slopes	4.5	0.10%
GfB	Glenford silt loam, 3 to 8 percent slopes	78.5	1.00%
GfC	Glenford silt loam, 8 to 15 percent slopes	45.5	0.60%
GuB	Guernsey silty clay loam, 3 to 8 percent slopes	5	0.10%
HaD	Hazleton channery loam, 15 to 25 percent slopes	47.1	0.60%
HeB	Hazleton loam, 3 to 8 percent slopes	8.2	0.10%
HeC	Hazleton loam, 8 to 15 percent slopes	155.6	2.00%
HeD	Hazleton loam, 15 to 25 percent slopes	343.9	4.30%
HeE	Hazleton loam, 25 to 40 percent slopes	111.9	1.40%
Ho	Holly silt loam, ponded	23.1	0.30%
LbB	Library Variant silt loam, 3 to 8 percent slopes	10.8	0.10%
MrD	Morristown shaly silty clay loam, 8 to 25 percent slopes	16.3	0.20%
Or	Orrville silt loam, occasionally flooded	169	2.10%
Tg	Tioga silt loam, occasionally flooded	21.6	0.30%
W	Water	3.6	0.00%
WhB	Wellston silt loam, 3 to 8 percent slopes	36.5	0.50%
WkC	Westmoreland silt loam, 8 to 15 percent slopes	182.7	2.30%
WkE	Westmoreland silt loam, 25 to 35 percent slopes	131.2	1.70%
WmC	Westmoreland-Coshocton silt loams, 8 to 15 percent slopes	1,134.00	14.30%
WmD	Westmoreland-Coshocton silt loams, 15 to 25 percent slopes	828.8	10.50%
Subtotals for Soil Survey Area		**4,878.30**	**61.50%**

1.2 Web Soil Survey for Carroll County, Ohio

2. Sampling Location: (A) Mine spoil hill with a star indicating the location of lysimeters, (B) Mine spoil Lysimeters, (C) Highwall hill with a start indicating the location of lysimeters, (D) Highwall Lysimeters

3. Weights of MS and HW Core after drying

MS	Bulk Soil + Ziplock bag (g)	HW	Bulk Soil + Ziplock bag (g)
0-10	356.64	0-10	406.22
10-20	460.6	10-20	562.77
20-30	514.2	20-30	507.47
30-40	459.06	30-40	550.46
40-50	447.43	40-50	562.15
50-60	360.91	50-60	590.73
60-70	371.62	60-70	541.52
70-80	407.5	70-80	530.75
80-90	430.17	80-90	541.52
90-100	420.27	90-100	550.97
100-110	473.99	100-110	345.19
110-120	437.96	110-120	470.25
Total	5140.35	**Total**	6160.00

4. XRD Diffractograms

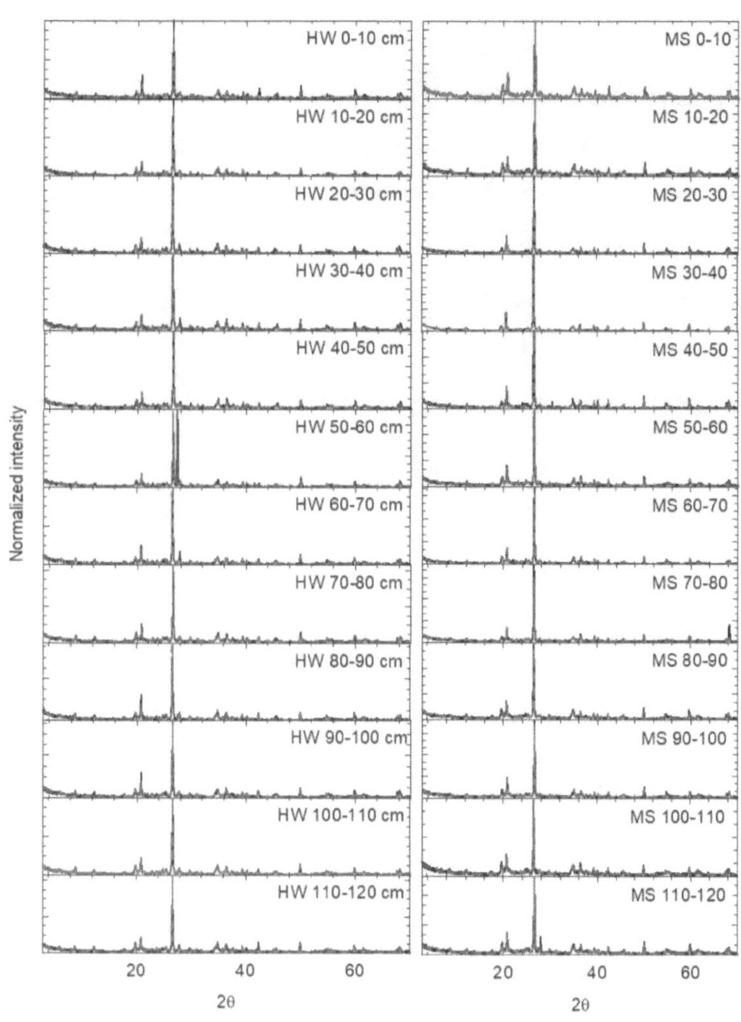

5. Field Data

5.1 Field Data for HW 10 cm

"X" indicates there was not enough water to extract for testing

A, B, C indicates which vial the water was stored

Day indicates when the sample was taken post-installation of the lysimeter

Sample	Date	Site	Depth (cm)	pH	EC (µS/cm)	Temperature (°C)	DO (mg/L)	A	B	C	Day
004	5/20/2015	HW	10	6.80	1700	12.4	8.70				7
012	5/27/2015	HW	10	6.46	1781	X	6.42				14
017	6/1/2015	HW	10	5.90	1428	14.3	7.93				19
028	6/4/2015	HW	10	7.17	1287	21.0	X		X		22
035	6/8/2015	HW	10	6.71	1055	22.7	X				26
044	6/11/2015	HW	10	6.50	854	24.2	5.59				29
052	6/15/2015	HW	10	6.62	612	24.4	X				33
060	6/18/2015	HW	10	6.18	473	22.5	5.97				36
068	6/22/2015	HW	10	6.21	202	22.3	4.64				40
076	6/25/2015	HW	10	6.22	479	20.7	5.35				43
084	6/29/2015	HW	10	5.85	420	18.8	5.33				47
092	7/2/2025	HW	10	6.20	414	21.5	5.27				50
100	7/6/2015	HW	10	6.72	350	24.1	X				54
108	7/9/2015	HW	10	6.65	122	23.3	X				57
116	7/28/2015	HW	10	6.49	370	25.5	3.23				76
124	8/6/2015	HW	10	6.58	346	20.3	5.27				85
132	8/13/2015	HW	10	X	X	X	X	X	X	X	92
140	8/27/2015	HW	10	X	X	X	X	X	X	X	106
148	10/2/2015	HW	10	X	X	X	X	X	X	X	142

5.2 Field Data for HW 40 cm

"X" indicates there was not enough water to extract for testing

A, B, C indicates which vial the water was stored

Day indicates when the sample was taken post-installation of the lysimeter

Sample	Date	Site	Depth (cm)	pH	EC (µS/cm)	Temperature (°C)	DO (mg/L)	A	B	C	Day
003	5/20/2015	HW	40	6.35	230	14.3	6.70				7
011	5/27/2015	HW	40	5.62	1635	X	5.02				14
018	6/1/2015	HW	40	5.87	2083	16.0	7.57				19
026	6/4/2015	HW	40	5.87	1678	18.8	6.10				22
036	6/8/2015	HW	40	X	X	X	X	X	X	X	26
043	6/11/2015	HW	40	X	X	X	X	X	X	X	29
050	6/15/2015	HW	40	5.49	747	23.0	X				33
058	6/18/2015	HW	40	5.86	664	21.9	3.00				36
066	6/22/2015	HW	40	5.70	667	21.7	4.05				40
074	6/25/2015	HW	40	5.52	562	20.7	4.05				43
082	6/29/2015	HW	40	5.87	539	18.8	4.76				47
090	7/2/2025	HW	40	5.77	437	19.6	4.49				50
098	7/6/2015	HW	40	6.01	441	20.9	4.11				54
106	7/9/2015	HW	40	6.02	408	21.2	3.05				57
114	7/28/2015	HW	40	7.09	355	24.7	3.48				76
122	8/6/2015	HW	40	6.01	262	20.4	4.05				85
130	8/13/2015	HW	40	X	X	X	X	X	X	X	92
138	8/27/2015	HW	40	X	X	X	X	X	X	X	106
146	10/2/2015	HW	40	X	X	X	X	X	X	X	142

5.3 Field Data for HW 80 cm

"X" indicates there was not enough water to extract for testing

A, B, C indicates which vial the water was stored

Day indicates when the sample was taken post-installation of the lysimeter

Sample	Date	Site	Depth (cm)	pH	EC (µS/cm)	Temperature (°C)	DO (mg/L)	A	B	C	Day
002	5/20/2015	HW	80	5.81	1268	12.9	6.10				7
010	5/27/2015	HW	80	6.39	1850	X	5.63				14
019	6/1/2015	HW	80	X	X	X	X	X	X	X	19
027	6/4/2015	HW	80	6.69	1725	19.0	5.05				22
034	6/8/2015	HW	80	X	X	X	X	X	X	X	26
042	6/11/2015	HW	80	X	X	X	X	X	X	X	29
051	6/15/2015	HW	80	6.40	878	22.9	X				33
059	6/18/2015	HW	80	X	X	X	X	X	X	X	36
067	6/22/2015	HW	80	5.80	890	20.8	4.84				40
075	6/25/2015	HW	80	5.57	622	20.2	4.85				43
083	6/29/2015	HW	80	X	X	X	X	X	X	X	47
091	7/2/2025	HW	80	5.78	518	19.3	3.91				50
099	7/6/2015	HW	80	5.82	414	20.5	3.58				54
107	7/9/2015	HW	80	6.17	388	20.8	3.48				57
115	7/28/2015	HW	80	6.22	323	23.7	2.92				76
123	8/6/2015	HW	80	6.27	313	19.8	4.67				85
131	8/13/2015	HW	80	5.99	289	23.9	X	X	X	X	92
139	8/27/2015	HW	80	X	X	X	X	X	X	X	106
147	10/2/2015	HW	80	X	X	X	X	X	X	X	142

5.4 Field Data for HW 120 cm

"X" indicates there was not enough water to extract for testing

A, B, C indicates which vial the water was stored

Day indicates when the sample was taken post-installation of the lysimeter

Sample	Date	Site	Depth (cm)	pH	EC (µS/cm)	Temperature (°C)	DO (mg/L)	A	B	C	Day
001	5/20/2015	HW	120	5.89	878	13.3	5.70				7
009	5/27/2015	HW	120	5.83	1865	X	4.57				14
020	6/1/2015	HW	120	7.20	1871	16.1	6.23				19
025	6/4/2015	HW	120	6.13	1650	19.4	6.05				22
033	6/8/2015	HW	120	6.26	1029	21.5	5.32				26
041	6/11/2015	HW	120	6.67	1366	21.5	4.99				29
049	6/15/2015	HW	120	6.28	1309	22.0	X				33
057	6/18/2015	HW	120	5.84	421	21.8	5.49				36
065	6/22/2015	HW	120	5.80	432	21.0	3.75				40
073	6/25/2015	HW	120	6.50	417	20.0	4.67				43
081	6/29/2015	HW	120	5.85	390	19.2	4.90				47
089	7/2/2025	HW	120	5.64	342	19.5	4.48				50
097	7/6/2015	HW	120	7.37	341	20.8	4.20				54
105	7/9/2015	HW	120	5.75	326	20.6	3.55				57
113	7/28/2015	HW	120	7.11	319	22.9	3.18				76
121	8/6/2015	HW	120	6.09	299	20.3	4.76				85
129	8/13/2015	HW	120	5.94	315	24.8	X				92
137	8/27/2015	HW	120	X	X	X	X	X	X	X	106
145	10/2/2015	HW	120	X	X	X	X	X	X	X	142

5.5 Field Data for MS 10 cm

"X" indicates there was not enough water to extract for testing

A, B, C indicates which vial the water was stored

Day indicates when the sample was taken post-installation of the lysimeter

Sample	Date	Site	Depth (cm)	pH	EC (µS/cm)	Temperature (°C)	DO (mg/L)	A	B	C	Day
008	5/20/2015	MS	10	6.74	1578	15.1	X				14
016	5/27/2015	MS	10	6.33	1159	X	X				19
024	6/1/2015	MS	10	X	X	X	X	X	X	X	22
032	6/4/2015	MS	10	X	X	X	X	X	X	X	26
040	6/8/2015	MS	10	6.41	545	22.0	X		X		26
048	6/11/2015	MS	10	6.08	1084	24.4	X				29
056	6/15/2015	MS	10	X	X	X	X	X	X	X	33
064	6/18/2015	MS	10	X	X	X	X	X	X	X	36
072	6/22/2015	MS	10	X	X	X	X	X	X	X	40
080	6/25/2015	MS	10	5.15	492	21.7	5.15				43
088	6/29/2015	MS	10	6.49	460	19.3	4.40				47
096	7/2/2025	MS	10	6.60	755	20.0	X				50
104	7/6/2015	MS	10	5.79	402	23.8	4.52	X	X	X	54
112	7/9/2015	MS	10	X	X	X	X	X	X	X	57
120	7/28/2015	MS	10	7.06	425	25.0	X				76
128	8/6/2015	MS	10	X	X	X	X	X	X	X	85
136	8/13/2015	MS	10	X	X	X	X	X	X	X	92
144	8/27/2015	MS	10	X	X	X	X	X	X	X	106
152	10/2/2015	MS	10	X	X	X	X	X	X	X	142

5.6 Field Data for MS 40 cm

"X" indicates there was not enough water to extract for testing

A, B, C indicates which vial the water was stored

Day indicates when the sample was taken post-installation of the lysimeter

Sample	Date	Site	Depth (cm)	pH	EC (µS/cm)	Temperature (°C)	DO (mg/L)	A	B	C	Day
007	5/20/2015	MS	40	6.18	276	14.3	4.00				7
015	5/27/2015	MS	40	5.81	159	X	5.27				14
023	6/1/2015	MS	40	5.49	323	16.2	6.67				19
031	6/4/2015	MS	40	5.33	635	21.1	6.31				22
039	6/8/2015	MS	40	5.54	582	21.1	X				26
047	6/11/2015	MS	40	5.41	245	23.8	5.36				29
055	6/15/2015	MS	40	5.38	382	25.2	X				33
063	6/18/2015	MS	40	4.78	440	21.1	3.98				36
071	6/22/2015	MS	40	4.89	535	21.1	4.35				40
079	6/25/2015	MS	40	4.13	523	20.0	4.13				43
087	6/29/2015	MS	40	5.32	256	19.5	4.40				47
095	7/2/2025	MS	40	5.44	432	19.5	3.27				50
103	7/6/2015	MS	40	5.23	381	22.0	X				54
111	7/9/2015	MS	40	X	X	X	X	X	X	X	57
119	7/28/2015	MS	40	7.06	316	24.5	2.41				76
127	8/6/2015	MS	40	5.39	239	19.9	3.87				85
135	8/13/2015	MS	40	X	X	X	X	X	X	X	92
143	8/27/2015	MS	40	X	X	X	X	X	X	X	106
151	10/2/2015	MS	40	X	X	X	X	X	X	X	142

5.7 Field Data for MS 80 cm

"X" indicates there was not enough water to extract for testing

A, B, C indicates which vial the water was stored

Day indicates when the sample was taken post-installation of the lysimeter

Sample	Date	Site	Depth (cm)	pH	EC (µS/cm)	Temperature (ºC)	DO (mg/L)	A	B	C	Day
006	5/20/2015	MS	80	5.76	233	14.4	6.20				7
014	5/27/2015	MS	80	5.11	691	X	3.34				14
022	6/1/2015	MS	80	5.85	620	16.8	X	X	X	X	19
030	6/4/2015	MS	80	5.20	582	19.5	3.51				22
038	6/8/2015	MS	80	5.55	773	20.3	4.11				26
046	6/11/2015	MS	80	5.09	683	24.4	3.71				29
054	6/15/2015	MS	80	5.22	629	21.7	X				33
062	6/18/2015	MS	80	5.08	569	20.0	3.27				36
070	6/22/2015	MS	80	5.00	550	19.7	3.08				40
078	6/25/2015	MS	80	5.04	363	20.0	3.20				43
086	6/29/2015	MS	80	5.23	430	18.4	3.90				47
094	7/2/2025	MS	80	5.55	192	20.7	3.86				50
102	7/6/2015	MS	80	5.24	333	21.7	2.38				54
110	7/9/2015	MS	80	5.45	160	22.1	2.65				57
118	7/28/2015	MS	80	7.07	297	24.4	2.00				76
126	8/6/2015	MS	80	5.35	303	19.3	3.45				85
134	8/13/2015	MS	80	4.97	251	21.3	3.20				92
142	8/27/2015	MS	80	X	X	X	X	X	X	X	106
150	10/2/2015	MS	80	X	X	X	X	X	X	X	142

5.8 Field Data for MS 120 cm

"X" indicates there was not enough water to extract for testing

A, B, C indicates which vial the water was stored

Day indicates when the sample was taken post-installation of the lysimeter

Sample	Date	Site	Depth (cm)	pH	EC (µS/cm)	Temperature (°C)	DO (mg/L)	A	B	C	Day
005	5/20/2015	MS	120	5.45	1010	12.6	4.00				7
013	5/27/2015	MS	120	5.24	1167	X	2.81				14
021	6/1/2015	MS	120	5.35	1237	15.4	3.71				19
029	6/4/2015	MS	120	6.61	904	23.9	X	X	X	X	22
037	6/8/2015	MS	120	5.91	638	21.0	X				26
045	6/11/2015	MS	120	5.69	356	23.7	4.86				29
053	6/15/2015	MS	120	5.83	798	24.0	X				33
061	6/18/2015	MS	120	5.89	726	21.5	X				36
069	6/22/2015	MS	120	5.63	600	21.1	5.00				40
077	6/25/2015	MS	120	5.63	664	20.2	X		X		43
085	6/29/2015	MS	120	X	X	X	X	X	X	X	47
093	7/2/2025	MS	120	X	X	X	X	X	X	X	50
101	7/6/2015	MS	120	X	X	X	X	X	X	X	54
109	7/9/2015	MS	120	X	X	X	X	X	X	X	57
117	7/28/2015	MS	120	X	X	X	X	X	X	X	76
125	8/6/2015	MS	120	6.24	606	22.4	X				85
133	8/13/2015	MS	120	X	X	X	X	X	X	X	92
141	8/27/2015	MS	120	X	X	X	X	X	X	X	106
149	10/2/2015	MS	120	X	X	X	X	X	X	X	142

6. Ion Chromatography

6.1 Fluoride IC concentrations for MS 10 cm, MS 40 cm, MS 80 cm, and MS 120 Cm

MS 10

Days Since Installation	Concentration mg/L	Concentration mol/L	Concentration umol/L
7	4.61	2.42E-04	242.46
14	8.90	4.69E-04	468.68
26	0.58	3.05E-05	30.48
29	0.67	3.55E-05	35.48
43	0.87	4.58E-05	45.79
47	0.79	4.16E-05	41.57
50	0.72	3.80E-05	38.02
		Average	**128.93**
		Error Bars	**168.10**

MS 40

Days Since Installation	Concentration mg/L	Concentration mol/L	Concentration umol/L
7	4.57	2.41E-04	240.53
14	4.45	2.34E-04	234.27
19	0.25	1.29E-05	12.91
22	0.45	2.38E-05	23.84
26	0.50	2.65E-05	26.47
29	0.51	2.70E-05	26.96
33	0.75	3.95E-05	39.54
36	0.48	2.51E-05	25.11
40	0.63	3.33E-05	33.28
43	0.60	3.16E-05	31.58
47	0.49	2.58E-05	25.85
50	0.49	2.57E-05	25.67
85	0.72	3.81E-05	38.13
		Average	**60.32**
		Error Bars	**78.89**

**MS
80**

Days Since Installation	Concentration mg/L	Concentration mol/L	Concentration umol/L
7	12.74	6.70E-04	670.46
14	4.77	2.51E-04	251.11
22	0.25	1.32E-05	13.17
26	0.24	1.27E-05	12.74
29	0.25	1.30E-05	12.95
33	0.45	2.38E-05	23.85
36	0.46	2.45E-05	24.45
40	0.45	2.37E-05	23.69
43	0.42	2.21E-05	22.11
47	0.44	2.31E-05	23.07
50	0.44	2.33E-05	23.28
85	0.50	2.64E-05	26.45
		Average	93.94
		Error Bars	193.34

**MS
120**

Days Since Installation	Concentration mg/L	Concentration mol/L	Concentration umol/L
7	8.49	4.47E-04	446.89
14	6.27	3.30E-04	329.86
19	0.26	1.39E-05	13.91
26	0.36	1.91E-05	19.11
29	0.36	1.88E-05	18.75
33	0.62	3.25E-05	32.47
36	0.64	3.34E-05	33.43
40	0.58	3.05E-05	30.53
43	0.55	2.90E-05	28.95
85	0.39	2.05E-05	20.53
		Average	97.44
		Error Bars	155.93

6.2 Fluoride IC concentrations for HW 10 cm, HW 40 cm, HW 80 cm, and HW 120 Cm

HW 10

Days Since Installation	Concentration mg/L	Concentration mol/L	Concentration umol/L
7	4.49	2.36E-04	236.39
14	4.70	2.48E-04	247.63
19	0.21	1.11E-05	11.14
22	0.27	1.41E-05	14.09
26	0.31	1.63E-05	16.25
29	0.32	1.68E-05	16.79
33	0.55	2.88E-05	28.78
36	0.64	3.35E-05	33.46
40	0.53	2.79E-05	27.90
43	0.54	2.84E-05	28.42
47	0.44	2.30E-05	22.97
50	0.45	2.38E-05	23.77
85	0.32	1.68E-05	16.84
		Average	55.73
		Error Bars	82.97

HW 40

Days Since Installation	Concentration mg/L	Concentration mol/L	Concentration umol/L
14	0.20	1.08E-05	10.76
33	0.33	1.72E-05	17.18
36	0.35	1.83E-05	18.29
43	0.32	1.68E-05	16.84
47	0.31	1.65E-05	16.49
50	0.32	1.66E-05	16.61
85	0.03	1.58E-06	1.58
		Average	13.96
		Error Bars	5.98

HW 80

Days Since Installation	Concentration mg/L	Concentration mol/L	Concentration umol/L
14	5.45	2.87E-04	286.67
22	0.35	1.84E-05	18.37
33	0.58	3.03E-05	30.32
40	0.60	3.15E-05	31.52
43	0.56	2.95E-05	29.48
50	0.53	2.80E-05	28.03
85	0.40	2.11E-05	21.05
92	0.50	2.62E-05	26.19
		Average	**58.95**
		Error Bars	**92.12**

HW 120

Days Since Installation	Concentration mg/L	Concentration mol/L	Concentration umol/L
26	0.17	8.69E-06	8.69
29	0.17	8.93E-06	8.93
33	0.40	2.12E-05	21.19
36	0.50	2.62E-05	26.23
40	0.33	1.74E-05	17.37
43	0.34	1.79E-05	17.90
47	0.32	1.69E-05	16.87
50	0.31	1.62E-05	16.19
85	0.34	1.79E-05	17.90
92	0.37	1.95E-05	19.53
		Average	**17.08**
		Error Bars	**5.22**

6.3 Chloride IC concentrations for MS 10 cm, MS 40 cm, MS 80 cm, and MS 120 Cm

MS 10

Days Since Installation	Concentration mg/L	Concentration mol/L	Concentration umol/L
7	46.90	1.36E-03	1361.19
14	23.11	6.71E-04	670.91
26	7.58	2.20E-04	220.01
29	7.10	2.06E-04	206.11
43	5.28	1.53E-04	153.31
47	4.24	1.23E-04	123.04
50	3.61	1.05E-04	104.75
		Average	405.62
		Error Bars	**464.00**

MS 40

Days Since Installation	Concentration mg/L	Concentration mol/L	Concentration umol/L
7	49.22	1.43E-03	1428.54
14	21.38	6.21E-04	620.60
19	7.60	2.21E-04	220.65
22	4.92	1.43E-04	142.82
26	3.94	1.14E-04	114.48
29	4.26	1.24E-04	123.54
33	3.88	1.13E-04	112.68
36	6.82	1.98E-04	197.85
40	4.36	1.27E-04	126.51
43	4.18	1.21E-04	121.26
47	3.28	9.53E-05	95.29
50	3.27	9.49E-05	94.90
85	2.0845997	6.05E-05	60.51
		Average	266.13
		Error Bars	**377.15**

MS

80

Days Since Installation	Concentration mg/L	Concentration mol/L	Concentration umol/L
7	36.95	1.07E-03	1072.41
14	5.21	1.51E-04	151.25
22	3.84	1.11E-04	111.38
26	3.62	1.05E-04	105.01
29	3.60	1.05E-04	104.53
33	4.04	1.17E-04	117.25
36	3.93	1.14E-04	113.93
40	4.14	1.20E-04	120.21
43	3.82	1.11E-04	110.76
47	3.16	9.18E-05	91.81
50	3.05	8.86E-05	88.57
85	1.91744838	5.57E-05	55.65
92	1.938478504	5.63E-05	56.26
		Average	176.85
		Error Bars	**270.29**

MS

120

Days Since Installation	Concentration mg/L	Concentration mol/L	Concentration umol/L
7	20.18	5.86E-04	585.62
14	8.26	2.40E-04	239.88
19	5.81	1.69E-04	168.59
26	4.61	1.34E-04	133.68
29	4.07	1.18E-04	118.05
33	4.97	1.44E-04	144.37
36	5.37	1.56E-04	155.82
40	5.16	1.50E-04	149.86
43	4.64	1.35E-04	134.55
85	2.982139495	8.66E-05	86.56
		Average	191.70
		Error Bars	**143.92**

6.4 Chloride IC concentrations for HW 10 cm, HW 40 cm, HW 80 cm, and HW 120 Cm

HW 10	Days Since Installation	Concentration mg/L	Concentration mol/L	Concentration umol/L
	7	47.59	1.38E-03	1381.37
	14	14.86	4.31E-04	431.19
	19	7.05	2.05E-04	204.67
	22	5.18	1.50E-04	150.38
	26	4.16	1.21E-04	120.86
	29	3.53	1.03E-04	102.59
	33	3.76	1.09E-04	109.22
	36	6.05	1.76E-04	175.57
	40	2.97	8.62E-05	86.25
	43	3.36	9.77E-05	97.66
	47	2.48	7.20E-05	72.00
	50	2.41	6.99E-05	69.88
	85	1.623213122	4.71E-05	47.11
			Average	234.52
			Error Bars	**358.33**

HW 40	Days Since Installation	Concentration mg/L	Concentration mol/L	Concentration umol/L
	7	87.72	2.55E-03	2546.15
	14	19.03	5.52E-04	552.28
	19	12.40	3.60E-04	359.96
	22	7.87	2.28E-04	228.44
	33	6.20	1.80E-04	179.92
	36	4.22	1.23E-04	122.50
	40	4.21	1.22E-04	122.17
	43	3.89	1.13E-04	112.87
	47	2.97	8.62E-05	86.23
	50	2.64	7.66E-05	76.57
	85	1.19	3.46E-05	34.64
			Average	401.98
			Error Bars	726.76

HW 80

Days Since Installation	Concentration mg/L	Concentration mol/L	Concentration umol/L
7	10.78	3.13E-04	312.75
14	13.27	3.85E-04	385.30
22	9.43	2.74E-04	273.65
33	7.96	2.31E-04	230.99
40	7.13	2.07E-04	206.92
43	6.15	1.78E-04	178.36
50	5.04	1.46E-04	146.35
85	3.286812687	9.54E-05	95.40
		Average	228.72
		Error Bars	93.48

HW 120

Days Since Installation	Concentration mg/L	Concentration mol/L	Concentration umol/L
7	30.66	8.90E-04	889.87
14	13.09	3.80E-04	379.81
19	10.25	2.97E-04	297.37
22	7.23	2.10E-04	209.84
26	6.18	1.79E-04	179.30
29	5.66	1.64E-04	164.36
33	5.95	1.73E-04	172.60
36	5.19	1.51E-04	150.56
40	3.67	1.07E-04	106.52
43	3.23	9.39E-05	93.87
47	2.23	6.46E-05	64.58
50	2.31	6.71E-05	67.09
85	1.30009049	3.77E-05	37.74
92	1.285155159	3.73E-05	37.30
		Average	203.63
		Error Bars	220.37

6.5 Nitrite IC concentrations for MS 10 cm, MS 40 cm, MS 80 cm, and MS 120 Cm

MS 10

Days Since Installation	Concentration mg/L	Concentration mol/L	Concentration umol/L
40	6.90	1.50E-04	150.08
47	7.35	1.60E-04	159.74
50	6.84	1.49E-04	148.60
		Average	152.81
		Error Bars	6.05

MS 40

Days Since Installation	Concentration mg/L	Concentration mol/L	Concentration umol/L
7	3.62	7.87E-05	78.68
40	6.15	1.34E-04	133.60
43	3.21	6.98E-05	69.78
47	3.03	6.58E-05	65.83
50	3.64	7.92E-05	79.22
		Average	85.42
		Error Bars	27.54

MS 80

Days Since Installation	Concentration mg/L	Concentration mol/L	Concentration umol/L
7	5.38	1.17E-04	116.94
14	6.88	1.50E-04	149.61
33	7.49	1.63E-04	162.72
40	6.02	1.31E-04	130.89
43	7.63	1.66E-04	165.80
47	4.55	9.88E-05	98.82
50	5.26	1.14E-04	114.33
		Average	134.16
		Error Bars	25.84

MS

120

Days Since Installation	Concentration mg/L	Concentration mol/L	Concentration umol/L
7	4.65	1.01E-04	101.18
14	6.74	1.46E-04	146.50
19	6.84	1.49E-04	148.62
26	5.35	1.16E-04	116.20
29	6.85	1.49E-04	148.91
33	6.99	1.52E-04	151.98
36	7.39	1.61E-04	160.62
40	9.63	2.09E-04	209.31
43	6.09	1.32E-04	132.43
		Average	146.19
		Error Bars	30.32

6.6 Nitrite IC concentrations for HW 10 cm, HW 40 cm, HW 80 cm, and HW 120 Cm

HW

10

Days Since Installation	Concentration mg/L	Concentration mol/L	Concentration umol/L
40	4.36	9.49E-05	94.87
43	4.76	1.03E-04	103.45
47	3.52	7.65E-05	76.49
50	3.30	7.18E-05	71.77
		Average	86.65
		Error Bars	14.99

**HW
40**

Days Since Installation	Concentration mg/L	Concentration mol/L	Concentration umol/L
40	3.81	8.28E-05	82.77
43	2.61	5.68E-05	56.75
47	2.77	6.02E-05	60.17
50	2.75	5.97E-05	59.69
		Average	64.85
		Error Bars	12.04

**HW
80**

Days Since Installation	Concentration mg/L	Concentration mol/L	Concentration umol/L
40	3.16	6.87E-05	68.71
43	2.86	6.21E-05	62.06
50	2.90	6.31E-05	63.06
		Average	64.61
		Error Bars	3.59

**HW
120**

Days Since Installation	Concentration mg/L	Concentration mol/L	Concentration umol/L
40	3.00	6.53E-05	65.26
43	2.78	6.04E-05	60.35
47	2.71	5.88E-05	58.81
50	2.93	6.37E-05	63.68
		Average	62.03
		Error Bars	2.97

6.7 Sulfate IC concentrations for MS 10 cm, MS 40 cm, MS 80 cm, and MS 120 Cm

MS 10

Days Since Installation	Concentration mg/L	Concentration mol/L	Concentration umol/L
7	993.68	1.03E-02	10344.24
14	858.84	8.94E-03	8940.52
26	534.71	5.57E-03	5566.40
29	487.56	5.08E-03	5075.55
43	466.55	4.86E-03	4856.84
47	433.73	4.52E-03	4515.14
50	378.41	3.94E-03	3939.29
		Average	6176.85
		Error Bars	2452.92

MS 40

Days Since Installation	Concentration mg/L	Concentration mol/L	Concentration umol/L
7	52.31	5.45E-04	544.51
14	191.08	1.99E-03	1989.18
19	113.32	1.18E-03	1179.67
22	293.60	3.06E-03	3056.38
26	257.82	2.68E-03	2683.90
29	264.14	2.75E-03	2749.73
33	261.04	2.72E-03	2717.49
36	243.86	2.54E-03	2538.58
40	260.73	2.71E-03	2714.16
43	249.92	2.60E-03	2601.68
47	230.34	2.40E-03	2397.89
50	202.18	2.10E-03	2104.66
		Average	2273.15
		Error Bars	731.63

MS 80	Days Since Installation	Concentration mg/L	Concentration mol/L	Concentration umol/L
1	7	33.67	3.51E-04	350.50
2	14	390.15	4.06E-03	4061.50
3	22	328.96	3.42E-03	3424.49
4	26	304.20	3.17E-03	3166.77
5	29	295.22	3.07E-03	3073.21
6	33	389.96	4.06E-03	4059.55
7	36	329.35	3.43E-03	3428.53
8	40	260.06	2.71E-03	2707.26
9	43	231.27	2.41E-03	2407.57
10	47	203.49	2.12E-03	2118.37
11	50	173.70	1.81E-03	1808.18
12			Average	2782.36
			Error Bars	1085.34

MS 120	Days Since Installation	Concentration mg/L	Concentration mol/L	Concentration umol/L
	7	458.62	4.77E-03	4774.28
	14	533.01	5.55E-03	5548.68
	19	411.31	4.28E-03	4281.81
	26	366.62	3.82E-03	3816.52
	29	386.18	4.02E-03	4020.19
	33	444.95	4.63E-03	4631.94
	36	402.36	4.19E-03	4188.57
	40	336.71	3.51E-03	3505.14
	43	304.05	3.17E-03	3165.17
			Average	4214.70
			Error Bars	713.55

6.8 Sulfate IC concentrations for HW 10 cm, HW 40 cm, HW 80 cm, and HW 120 Cm

HW 10

Days Since Installation	Concentration mg/L	Concentration mol/L	Concentration umol/L
7	607.23	6.32E-03	6321.32
14	798.78	8.32E-03	8315.35
19	664.33	6.92E-03	6915.66
22	444.61	4.63E-03	4628.40
26	342.42	3.56E-03	3564.63
29	263.07	2.74E-03	2738.54
33	245.58	2.56E-03	2556.55
36	199.70	2.08E-03	2078.86
40	178.59	1.86E-03	1859.13
43	41.80	4.35E-04	435.11
47	129.01	1.34E-03	1342.99
50	115.72	1.20E-03	1204.68
		Average	3496.77
		Std Deviation	2515.20

HW 40

Days Since Installation	Concentration mg/L	Concentration mol/L	Concentration umol/L
7	61.58	6.41E-04	641.08
14	827.88	8.62E-03	8618.29
19	1027.44	1.07E-02	10695.73
22	747.64	7.78E-03	7782.96
33	610.09	6.35E-03	6351.03
36	336.51	3.50E-03	3503.11
40	285.95	2.98E-03	2976.73
43	273.30	2.85E-03	2845.02
47	213.49	2.22E-03	2222.42
50	154.33	1.61E-03	1606.57
		Average	4724.29
		Std Deviation	3392.02

HW 80

Days Since Installation	Concentration mg/L	Concentration mol/L	Concentration umol/L
7	585.78	6.10E-03	6098.03
14	988.00	1.03E-02	10285.17
22	758.10	7.89E-03	7891.88
33	571.53	5.95E-03	5949.64
40	385.96	4.02E-03	4017.91
43	270.94	2.82E-03	2820.50
50	205.27	2.14E-03	2136.86
		Average	5600.00
		Std Deviation	2880.85

HW 120

Days Since Installation	Concentration mg/L	Concentration mol/L	Concentration umol/L
7	374.39	3.90E-03	3897.45
14	931.72	9.70E-03	9699.22
19	896.44	9.33E-03	9332.03
22	770.82	8.02E-03	8024.28
26	685.84	7.14E-03	7139.66
29	618.59	6.44E-03	6439.54
33	597.82	6.22E-03	6223.29
36	289.90	3.02E-03	3017.84
40	196.42	2.04E-03	2044.72
43	327.45	3.41E-03	3408.72
47	143.81	1.50E-03	1497.02
50	142.74	1.49E-03	1485.96
		Average	5184.14
		Std Deviation	2996.44

6.9 Bromide IC concentrations for MS 10 cm, MS 40 cm, MS 80 cm, and MS 120 Cm

MS 10

Days Since Installation	Concentration mg/L	Concentration mol/L	Concentration umol/L
43	2.11	2.63E-05	26.34
		Average	26.34
		Error Bars	NA

MS 40

Days Since Installation	Concentration mg/L	Concentration mol/L	Concentration umol/L
7	0.90	1.12E-05	11.23
29	0.91	1.14E-05	11.43
50	35.42	4.43E-04	443.33
		Average	155.33

MS 80

Days Since Installation	Concentration mg/L	Concentration mol/L	Concentration umol/L
26	0.93	1.16E-05	11.65
40	2.04	2.55E-05	25.55
43	2.06	2.58E-05	25.82
47	1.63	2.04E-05	20.42
50	1.64	2.05E-05	20.50
		Average	20.79
		Error Bars	5.74

MS
120

Days Since Installation	Concentration mg/L	Concentration mol/L	Concentration umol/L
26	0.99	1.24E-05	12.35
29	0.98	1.23E-05	12.32
36	2.39	2.99E-05	29.92
43	2.04	2.55E-05	25.47
		Average	20.02
		Error Bars	9.06

6.10 Bromide IC concentrations for HW 10 cm, HW 40 cm, HW 80 cm, and HW 120 Cm

*Note: There was no bromide data for HW 10 cm

HW
40

Days Since Installation	Concentration mg/L	Concentration mol/L	Concentration umol/L
7	0.95	1.18E-05	11.85
36	56.75	7.10E-04	710.17
40	41.31	5.17E-04	516.94
50	32.98	4.13E-04	412.75
		Average	412.93
		Std Deviation	294.41

HW 80

Days Since Installation	Concentration mg/L	Concentration mol/L	Concentration umol/L
50	33.26	4.16E-04	416.28
		Average	416.28
		Std Deviation	NA

HW 120

Days Since Installation	Concentration mg/L	Concentration mol/L	Concentration umol/L
47	32.86	4.11E-04	411.21
		Average	411.21
		Std Deviation	NA

6.11 Phosphate IC concentrations for MS 10 cm, MS 40 cm, MS 80 cm, and MS 120 Cm

MS 10

Days Since Installation	Concentration mg/L	Concentration mol/L	Concentration umol/L
47	57.38	6.04E-04	604.19
		Average	604.19
		Std Deviation	NA

**MS
40**

Days Since Installation	Concentration mg/L	Concentration mol/L	Concentration umol/L
14	1.03	1.08E-05	10.84
22	20.62	2.17E-04	217.13
29	2.26	2.38E-05	23.79
36	4.23	4.46E-05	44.56
40	4.18	4.40E-05	43.98
43	4.08	4.29E-05	42.93
50	3.69	3.89E-05	38.90
		Average	60.30
		Error Bars	70.30

**MS
80**

Days Since Installation	Concentration mg/L	Concentration mol/L	Concentration umol/L
7	0.98	1.03E-05	10.33
26	2.05	2.16E-05	21.56
33	6.27	6.60E-05	66.00
43	4.20	4.42E-05	44.21
47	3.30	3.47E-05	34.74
50	3.41	3.59E-05	35.93
		Average	35.46
		Error Bars	19.15

**MS
120**

Days Since Installation	Concentration mg/L	Concentration mol/L	Concentration umol/L
19	1.15	1.21E-05	12.11
26	2.29	2.41E-05	24.08
29	2.49	2.63E-05	26.25
33	3.77	3.97E-05	39.72
36	4.48	4.71E-05	47.12
43	4.21	4.44E-05	44.37
		Average	32.28
		Error Bars	13.65

6.12 Phosphate IC concentrations for HW 10 cm, HW 40 cm, HW 80 cm, and HW 120 Cm

HW 10

Days Since Installation	Concentration mg/L	Concentration mol/L	Concentration umol/L
7	1.34	1.41E-05	14.11
14	1.18	1.24E-05	12.44
19	1.53	1.61E-05	16.08
22	1.59	1.67E-05	16.70
26	40.07	4.22E-04	421.95
29	36.09	3.80E-04	380.02
33	4.45	4.69E-05	46.86
36	5.43	5.71E-05	57.12
40	5.39	5.67E-05	56.74
43	5.01	5.27E-05	52.72
47	3.15	3.31E-05	33.13
50	3.12	3.29E-05	32.87
		Average	95.06
		Std Deviation	144.14

HW 40

Days Since Installation	Concentration mg/L	Concentration mol/L	Concentration umol/L
22	1.25	1.31E-05	13.13
33	3.69	3.89E-05	38.88
36	3.97	4.18E-05	41.82
40	0.62	6.54E-06	6.54
43	3.88	4.09E-05	40.87
47	3.38	3.56E-05	35.58
50	3.44	3.63E-05	36.25
		Average	30.44
		Std Deviation	14.38

HW 80	Days Since Installation	Concentration mg/L	Concentration mol/L	Concentration umol/L
	33	3.79	3.99E-05	39.93
	40	4.04	4.25E-05	42.51
	50	3.52	3.71E-05	37.11
			Average	39.85
			Std Deviation	2.70

HW 120	Days Since Installation	Concentration mg/L	Concentration mol/L	Concentration umol/L
	7	22.15	2.33E-04	233.22
	19	26.56	2.80E-04	279.69
	29	1.82	1.92E-05	19.20
	36	4.02	4.23E-05	42.28
	40	3.78	3.98E-05	39.83
	43	3.96	4.16E-05	41.65
	47	2.84	2.99E-05	29.87
	50	3.31	3.49E-05	34.90
			Average	90.08
			Std Deviation	103.71

6.13 Calibration curves for IC anions

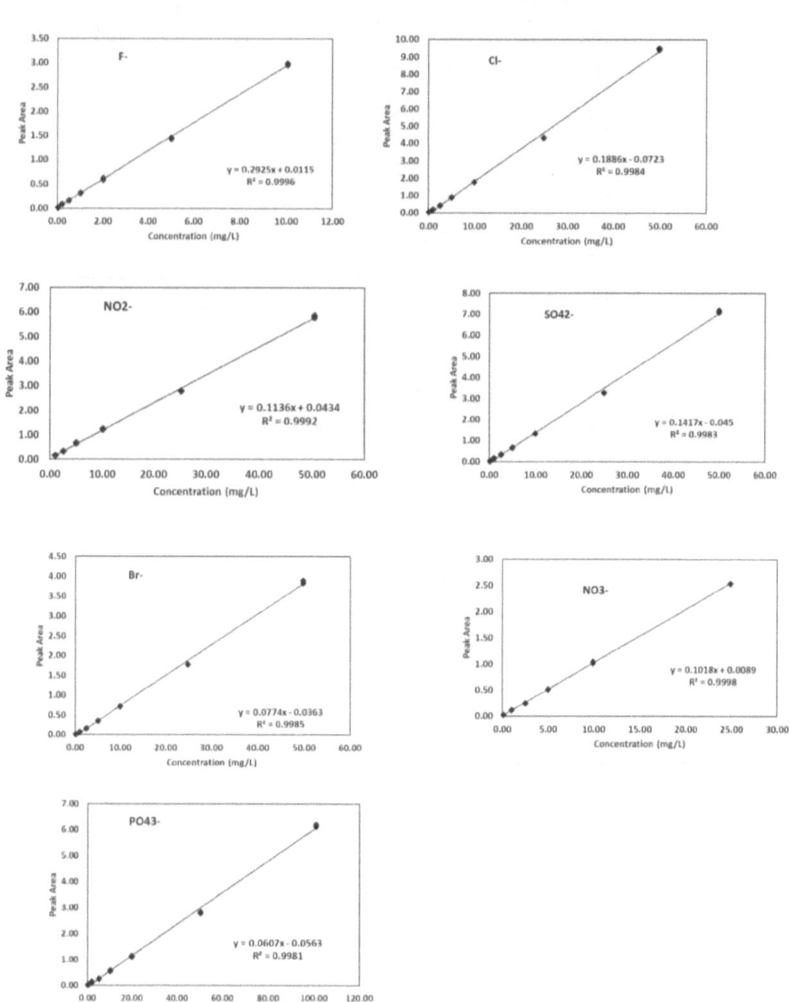

95

7. Dissolved Organic Carbon

7.1 Dissolved Organic Carbon HW Data

Site	Depth (cm)	Conc. mg/L	Error Bars	Depth (cm)	Conc. mg/L	Error Bars	Depth (cm)	Conc. mg/L	Error Bars	Depth (cm)	Conc. mg/L	Error Bars
HW	10	81.42	1.13	40	7.16	1.13	80	11.03	1.13	120	6.73	1.13
HW	10	71.49	1.13	40	10.16	1.13	80	9.10	1.13	120	9.49	0.43
HW	10	76.47	0.43	40	12.41	0.43	80	10.13	0.43	120	12.32	
HW	10	90.47	0.47	40	15.12	0.43	80	15.13		120	13.82	
HW	10	80.48	0.47	40	9.83	0.35	80	8.55		120	13.27	
HW	10	87.34	0.35	40	17.74	0.35	80	6.89		120	7.59	
HW	10	91.09	0.35	40	17.15	0.54	80	8.34	0.86	120	7.02	
HW	10	99.15	0.54	40	14.11	0.54	80	9.20	0.69	120	7.92	
HW	10	110.53	0.54	40	19.22	0.86	80	8.07	0.69	120	7.54	0.86
HW	10	99.31	0.86	40	17.52	0.86	80	6.89	3.68	120	7.79	0.86
HW	10	95.22	0.86	40	18.11	0.69	80	6.94	3.68	120	7.98	0.69
HW	10	102.68	0.69	40	17.05	0.69				120	9.02	0.69
HW	10	111.75	0.69	40	11.76	3.68				120	7.88	3.68
HW	10	87.11	3.68	40	10.84	3.68				120	7.73	3.68
HW	10	87.17	3.68							120	6.12	0.67
	10 cm Average: 91.45 mg/L			40 cm Average: 14.12 mg/L			80 cm Average: 9.12 mg/L			120 cm Average: 8.81 mg/L		
	10 cm Std Error 1.06 mg/L			40 cm Std Error: 3.78 mg/L			80 cm Std Error: 2.75 mg/L			120 cm Std Error: 2.28 mg/L		

7.2 Dissolved Organic Carbon MS Data

Site	Depth (cm)	Conc. mg/L	Error Bars	Depth (cm)	Conc. mg/L	Error Bars	Depth (cm)	Conc. mg/L	Error Bars	Depth (cm)	Conc. mg/L	Error Bars
MS	10	13.31	1.13	40	10.68	1.13	80	18.65	1.13	120	24.81	1.13
MS	10	9.27	1.13	40	8.45	1.13	80	23.46	1.13	120	37.97	1.13
MS	10	9.41	0.47	40	11.35	0.43	80	22.88	0.43	120	41.99	0.43
MS	10	22.06	0.54	40	8.52	0.43	80	26.23	0.47	120	30.09	
MS	10	28.02	0.86	40	9.05	0.47	80	25.89	0.47	120	25.95	
MS	10	26.26	0.86	40	8.01	0.47	80	19.37	0.35	120	19.89	
MS	10	10.56	3.68	40	8.21	0.35	80	20.46	0.35	120	24.31	
MS	10			40	10.20	0.35	80	24.74		120	26.26	
MS				40	18.06	0.54	80	23.59		120	12.45	3.68
MS				40	15.24	0.54	80	23.85	0.86			
MS				40	18.06	0.86	80	28.14	0.86			
MS				40	11.33	0.86	80	31.38	0.69			
MS				40	15.31	0.69	80	32.24	0.69			
MS				40	9.93	3.68	80	15.77	3.68			
MS				40	8.99	3.68	80	17.43	3.68			
MS							80	8.59	0.67			
	10 cm Average: 16.98 mg/L			40 cm Average: 11.43 mg/L			80 cm Average: 22.67 mg/L			120 cm Average: 27.08 mg/L		
	10 cm Std Error 6.42 mg/L			40 cm Std Error: 2.95 mg/L			80 cm Std Error: 5.67 mg/L			120 cm Std Error: 9.03 mg/L		

8. Particle Size Analysis

Depth (cm)	High Wall			Mine Spoil		
	clay	silt	sand	clay	silt	sand
0-10	4.7	54.1	41.2	7.2	47.8	45.0
10-20	4.6	51.9	43.5	7.1	44.4	48.4
20-30	3.2	37.6	59.2	5.4	45.0	49.6
30-40	3.6	41.4	55.0	5.4	43.3	51.3
40-50	4.6	48.6	46.8	5.4	42.3	52.3
50-60	4.6	51.2	44.2	5.1	38.0	56.9
60-70	4.7	47.9	47.4	5.6	35.7	58.7
70-80	4.8	48.6	46.5	6.2	39.7	54.0
80-90	4.1	43.2	52.7	7.2	44.5	48.3
90-100	4.5	47.0	48.5	6.5	42.2	51.2
100-110	4.4	48.8	46.8	7.4	40.4	52.3
110-120	4.8	52.6	42.6	6.7	38.5	54.7

9. XRD Analysis

	High Wall					Mine Spoil				
Depth (cm)	quartz	feldspar	kaolinite	muscovite	chlorite	quartz	feldspar	kaolinite	muscovite	chlorite
0-10	41	6.8	13.6	36.6	2.1	34	3.7	20.5	38.6	3.2
10-20	34.5	6.1	14.9	41.4	3.3	27.5	3.5	21.7	45	2.3
20-30	35.1	8.3	14.2	39.3	3	41.9	4.2	16.5	35.3	2.1
30-40	43.3	6.5	12.3	34.9	3	49	3.8	16.1	29.1	2.1
40-50	37.8	6.5	13.1	38.8	3.8	41.7	6.5	18.6	30.9	2.4
50-60	30.6	8.7	15.3	42.6	2.9	34.3	6.2	20	35.4	3.9
60-70	38	6.9	14.6	37	3.5	34.6	6	18.8	36.5	5
70-80	35.6	3.9	15.5	42.3	2.6	44	4	17.9	33.3	0.8
80-90	38.9	5.7	13	38.4	3.9	34.7	1.1	22.4	38.1	3.7
90-100	37.5	6	15.3	38.5	2.6	37.3	4.8	19.9	33.4	4.6
100-110	36.3	5.4	15	39.3	4	28.7	4.5	20.8	38.1	7.8
110-120	34.8	5.9	14.3	41.6	3.4	31.1	9.4	18.3	40	1.1

10. Loss On Ignition

10.1 MS LOI at 550°C

Table X

Depth	Weight of Crucible (g)	Weight of Crucible + Soil (g)	Soil Added (g)	1 g - Soil Added	Weight of Crucible + Soil After Heating (g)	Weight Lost (g)	Depth	% Lost
0-10	13.833	14.833	1.000	0.000	14.746	0.087	0-10	8.7
0-10	13.470	14.476	1.005	-0.005	14.373	0.102	0-10	10.73
0-10	13.442	14.445	1.003	-0.003	14.344	0.101	0-10	10.42
10-20	13.693	14.693	1.000	0.000	14.617	0.076	10-20	7.6
10-20	13.833	14.819	0.986	0.014	14.742	0.078	10-20	6.38
10-20	13.894	14.881	0.987	0.013	14.803	0.078	10-20	6.51
20-30	13.249	14.249	1.000	0.000	14.195	0.054	20-30	5.4
20-30	13.894	14.891	0.998	0.002	14.826	0.065	20-30	6.25
20-30	13.799	14.791	0.991	0.009	14.726	0.065	20-30	5.59
30-40	11.936	12.936	1.000	0.000	12.892	0.044	30-40	4.4
30-40	13.693	14.709	1.017	-0.017	14.655	0.055	30-40	7.12
30-40	13.894	14.885	0.992	0.008	14.832	0.054	30-40	4.53
40-50	12.750	13.750	1.000	0.000	13.696	0.054	40-50	5.4
40-50	13.470	14.465	0.994	0.006	14.399	0.066	40-50	6.03
40-50	13.877	14.875	0.997	0.003	14.808	0.066	40-50	6.35
50-60	11.517	12.517	1.000	0.000	12.393	0.124	50-60	12.4
50-60	11.517	12.545	1.028	-0.028	12.404	0.141	50-60	16.88
50-60	13.833	14.864	1.031	-0.031	14.723	0.141	50-60	17.15
60-70	13.443	14.443	1.000	0.000	14.333	0.110	60-70	11
60-70	11.937	12.929	0.992	0.008	12.812	0.117	60-70	10.94
60-70	13.945	14.942	0.997	0.003	14.824	0.118	60-70	11.44
70-80	13.661	14.661	1.000	0.000	14.572	0.089	70-80	8.9
70-80	13.249	14.223	0.975	0.025	14.128	0.095	70-80	6.99
70-80	12.750	13.747	0.998	0.002	13.648	0.099	70-80	9.68
80-90	13.877	14.877	1.000	0.000	14.761	0.116	80-90	11.6
80-90	13.833	14.844	1.011	-0.011	14.714	0.129	80-90	14
80-90	13.945	14.942	0.997	0.003	14.814	0.128	80-90	12.48
90-100	13.799	14.799	1.000	0.000	14.697	0.102	90-100	10.2
90-100	13.249	14.247	0.998	0.002	14.087	0.159	90-100	15.7
90-100	12.928	13.925	0.997	0.003	13.811	0.114	90-100	11.13
100-110	13.470	14.470	1.000	0.000	14.376	0.094	100-110	9.4
100-110	13.945	14.930	0.984	0.016	14.821	0.109	100-110	9.3
100-110	13.800	14.798	0.999	0.001	14.689	0.109	100-110	10.77
110-120	12.928	13.928	1.000	0.000	13.846	0.082	110-120	8.2
110-120	12.928	13.951	1.023	-0.023	13.856	0.096	110-120	11.9
110-120	13.662	14.669	1.007	-0.007	14.572	0.096	110-120	10.29

Loss On Ignition for Mine Tailing Hill (MT) at 550 °C

10.2 HW LOI at 550°C

Depth	Weight of Crucible (g)	Weight of Crucible + Soil (g)	Soil Added (g)	1 g - soil added	Weight of Crucible + Soil After Heating (g)	Weight Lost (g)	Depth	% Lost
0-10	13.833	14.833	1.000	0.000	14.746	0.087	0-10	8.7
0-10	12.928	13.929	1.002	-0.002	13.863	0.066	0-10	6.77
0-10	11.517	12.515	0.998	0.002	12.449	0.066	0-10	6.41
10-20	13.693	14.693	1.000	0.000	14.617	0.076	10-20	7.6
10-20	13.249	14.255	1.006	-0.005	14.190	0.065	10-20	7
10-20	11.517	12.506	0.989	0.011	12.442	0.064	10-20	5.26
20-30	13.249	14.249	1.000	0.000	14.195	0.054	20-30	5.4
20-30	11.937	12.945	1.009	-0.008	12.880	0.065	20-30	7.33
20-30	13.945	14.966	1.021	-0.021	14.900	0.066	20-30	8.72
30-40	11.936	12.936	1.000	0.000	12.892	0.044	30-40	4.4
30-40	13.693	14.687	0.995	0.005	14.633	0.054	30-40	4.9
30-40	13.470	14.452	0.982	0.018	14.398	0.053	30-40	3.52
40-50	12.750	13.750	1.000	0.000	13.696	0.054	40-50	5.4
40-50	13.662	14.663	1.002	-0.002	14.597	0.066	40-50	6.75
40-50	12.926	13.926	1.000	0.000	13.861	0.065	40-50	6.49
50-60	11.517	12.517	1.000	0.000	12.393	0.124	50-60	12.4
50-60	13.800	14.770	0.970	0.030	14.705	0.065	50-60	3.52
50-60	13.442	14.470	1.028	-0.028	14.401	0.069	50-60	9.67
60-70	13.443	14.443	1.000	0.000	14.333	0.110	60-70	11
60-70	13.799	14.803	1.003	-0.003	14.742	0.061	60-70	6.47
60-70	13.877	14.864	0.987	0.013	14.805	0.060	60-70	4.72
70-80	13.661	14.661	1.000	0.000	14.572	0.089	70-80	8.9
70-80	12.750	13.748	0.998	0.002	13.685	0.063	70-80	6.12
70-80	11.516	12.526	1.009	-0.009	12.463	0.063	70-80	7.21
80-90	13.877	14.877	1.000	0.000	14.761	0.116	80-90	11.6
80-90	13.249	14.250	1.001	-0.001	14.189	0.061	80-90	6.18
80-90	13.442	14.432	0.989	0.011	14.371	0.061	80-90	5.01
90-100	13.799	14.799	1.000	0.000	14.697	0.102	90-100	10.2
90-100	13.693	14.695	1.002	-0.002	14.634	0.061	90-100	6.26
90-100	11.936	12.930	0.994	0.006	12.871	0.059	90-100	5.31
100-110	13.470	14.470	1.000	0.000	14.376	0.094	100-110	9.4
100-110	11.937	12.942	1.005	-0.005	12.881	0.061	100-110	6.6
100-110	13.878	14.895	1.017	-0.017	14.833	0.061	100-110	7.82
110-120	12.928	13.928	1.000	0.000	13.846	0.082	110-120	8.2
110-120	13.877	14.871	0.993	0.007	14.811	0.059	110-120	5.26
110-120	13.693	14.695	1.002	-0.002	14.635	0.060	110-120	6.18

Loss On Iginition for High Wall (HW) at 550 °C

11. Sequential Extraction

11.1: Average concentrations of aluminum during sequential extraction procedure.

Aluminum	Solution 1: Exchangeables mmol kg^{-1}	Solution 2: Bound to Carbonates mmol kg^{-1}	Solution 3: Reducibles mmol kg^{-1}	Solution 4: Oxidizables mmol kg^{-1}
HW				
0-10	4.66	1.91	150.88	40.38
10-20	1.33	1.03	153.62	39.17
20-30	0.58	0.81	144.20	42.79
30-40	0.25	0.75	140.89	39.91
40-50	0.16	0.41	129.81	35.91
50-60	0.15	0.36	137.07	42.55
60-70	1.36	1.09	135.09	45.43
70-80	2.39	1.08	126.73	35.72
80-90	2.38	1.01	114.56	37.78
90-100	8.31	1.56	147.83	49.56
100-110	0.25	0.75	120.59	36.97
110-120	0.36	0.81	95.46	44.63
Total concentration mmol kg^{-1}	22.18	11.56	1596.74	490.79
Total mmol kg^{-1} Al in profile	2121.27			
MS				
0-10	1.37	1.23	179.99	49.30
10-20	1.76	1.34	156.24	44.69
20-30	1.52	1.31	172.56	43.32
30-40	0.92	1.25	190.78	40.24
40-50	1.03	1.09	204.19	39.69
50-60	3.32	1.47	178.35	52.93
60-70	1.32	1.02	193.97	51.16
70-80	0.40	0.89	215.70	47.45
80-90	0.93	1.11	235.72	51.82
90-100	4.01	1.62	215.52	46.94
100-110	3.03	1.48	175.56	41.77
110-120	0.85	0.94	160.36	48.65
Total concentration mmol kg-1	20.46	14.75	2278.93	557.97
Total mmol kg^{-1} Al in profile	2872.12			

11.2: Average concentrations of manganese during sequential extraction procedure.

Manganese	Solution 1: Exchangeables mmol kg^{-1}	Solution 2: Bound to Carbonates mmol kg^{-1}	Solution 3: Reducibles mmol kg^{-1}	Solution 4: Oxidizables mmol kg^{-1}
HW				
0-10	0.22	0.09	2.09	–
10-20	–	0.04	2.08	–
20-30	–	0.05	2.06	–
30-40	–	0.04	1.45	–
40-50	–	0.04	1.31	–
50-60	–	0.03	1.32	–
60-70	–	0.05	1.31	–
70-80	–	0.04	1.79	–
80-90	–	0.03	1.46	–
90-100	–	0.07	2.55	–
100-110	–	0.05	2.04	–
110-120	–	0.05	1.35	–
Total concentration mmol kg^{-1}	0.22	0.60	20.81	0.00
Total mmol kg^{-1} Mn in profile	21.62			
MS				
0-10	1.52	0.38	15.79	–
10-20	0.60	0.15	13.37	–
20-30	0.49	0.11	7.48	–
30-40	0.36	0.10	7.62	–
40-50	0.31	0.08	7.58	–
50-60	0.28	0.09	7.64	–
60-70	0.46	0.13	8.56	–
70-80	0.35	0.16	53.97	–
80-90	0.32	0.13	53.16	–
90-100	0.37	0.10	60.05	–
100-110	0.54	0.13	15.93	–
110-120	0.40	0.12	13.50	–
Total concentration mmol kg-1	5.98	1.70	264.65	0.00
Total mmol kg^{-1} Mn in profile	272.33			

11.3: Average concentrations of iron during sequential extraction procedure.

Iron	Solution 1: Exchangeables mmol kg^{-1}	Solution 2: Bound to Carbonates mmol kg^{-1}	Solution 3: Reducibles mmol kg^{-1}	Solution 4: Oxidizables mmol kg^{-1}
HW				
0-10	–	0.37	608.9	7.33
10-20	–	0.05	622.7	4.59
20-30	–	0.05	584.7	6.12
30-40	–	0.04	587.9	6.55
40-50	–	0.02	545.8	6.40
50-60	–	0.01	559.0	8.12
60-70	–	0.07	702.0	7.10
70-80	–	0.12	603.4	6.37
80-90	–	0.07	504.7	5.24
90-100	–	0.21	820.5	24.88
100-110	–	0.06	636.9	4.40
110-120	–	0.09	410.5	5.87
Total concentration mmol kg^{-1}	0.00	1.16	7187.04	92.95
Total mmol kg^{-1} Fe in profile	7281.15			
MS				
0-10	–	0.16	663.6	10.53
10-20	–	0.13	571.7	6.28
20-30	–	0.08	344.0	4.89
30-40	–	0.14	539.9	4.71
40-50	–	0.08	535.0	6.21
50-60	–	0.22	537.3	20.85
60-70	–	0.08	717.3	15.79
70-80	–	0.04	1552.2	19.45
80-90	–	0.28	1518.5	16.46
90-100	–	0.40	1701.1	12.82
100-110	–	0.39	746.9	17.88
110-120	–	0.09	780.0	14.46
Total concentration mmol kg-1	0.00	2.07	10207.44	150.33
Total mmol kg^{-1} Fe in profile	10359.84			

12. Dissolved Oxygen

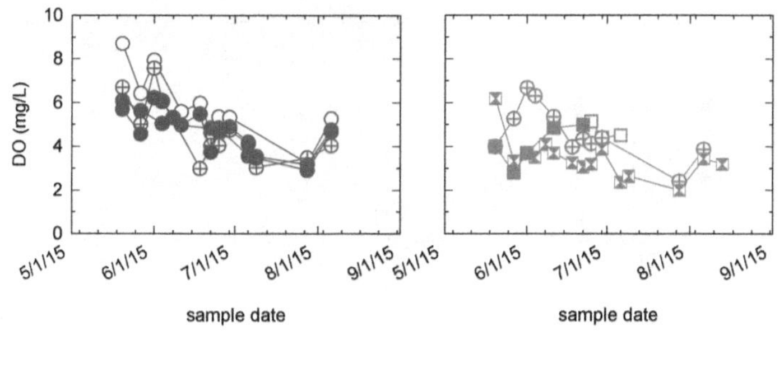

13. pH

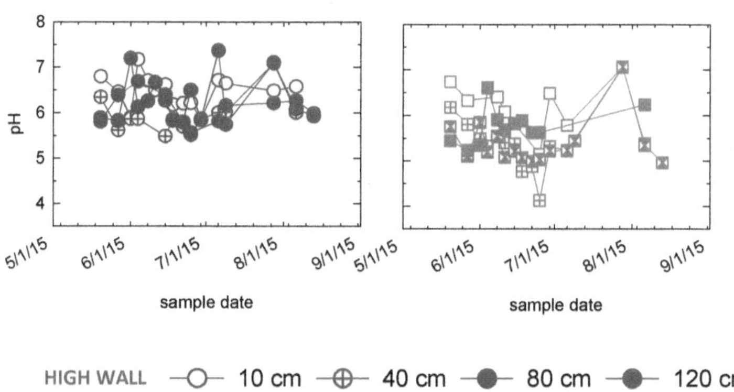

14. Electrical Conductivity

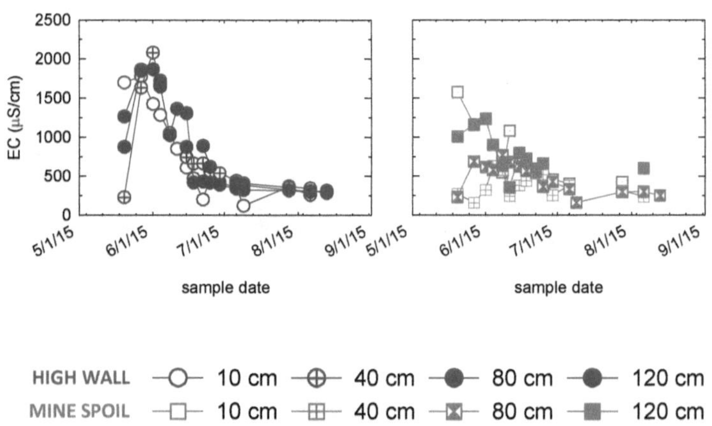

15. Temperature of soil Pore Water

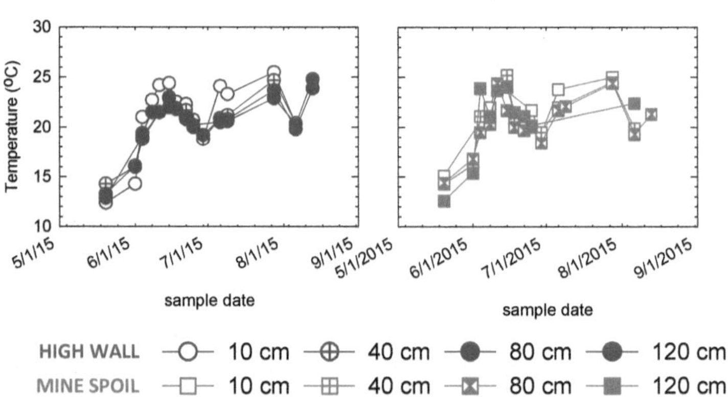

106

www.ingramcontent.com/pod-product-compliance
Lightning Source LLC
LaVergne TN
LVHW041712190726
843493LV00007B/2052